FUNCTIONAL ANALYSIS

Simplify before Automating

FUNCTIONAL ANALYSIS

Simplify before Automating

Eugene J. Wittry

VNR VAN NOSTRAND REINHOLD
New York

Library of Congress Catalog Card Number 90-47491
ISBN 0-442-00583-0

Printed in the United States of America.

Van Nostrand Reinhold
115 Fifth Avenue
New York, New York 10003

Chapman and Hall
2–6 Boundary Row
London, SE1 8HN, England

Thomas Nelson Australia
102 Dodds Street
South Melbourne 3205
Victoria, Australia

Nelson Canada
1120 Birchmount Road
Scarborough, Ontario M1K 5G4, Canada

16 15 14 13 12 11 10 9 8 7 6 5 4 3 2 1

Library of Congress Cataloging-in-Publication Data

Wittry, Eugene J.
Functional analysis : simplify before automating / Eugene J. Wittry.
p. cm.
Includes index.
ISBN 0-442-00583-0
1. Management information systems. I. Title.
T58.6.W566 1991
658.4'038—dc20 90-47491
CIP

To Al Hagemann, a very dear friend and supporter. His personal success in business is the source of the Gardentool examples in this book.

Contents

Preface

This book is a cookbook. It is a "how-to" set of instructions for functional analysis (FA). So what is FA? It is a method for analyzing the information system needs of a business.

That is nice. So why do we need one more technique when there are so many of them? There are a number of techniques on the market for analyzing the system needs of a business. These include SISP (strategic information system planning), BAA (business area analysis), BSP (business systems planning), CSFs (critical success factors), and a whole host of other techniques. What need is there for one more method for doing the same job?

FA is different from these other techniques in one significant way: It does not start with the presumption that the answer to an information system problem must be a new computer system. The aim of FA is to simplify business practices—not to automate them. Simplification must come first. If automation is needed, it can be done—*after* the need for automation is proved. Automation is often needed to improve business practices, but it is not always needed. FA is a method for ensuring that the right tool is used for the right job.

For a number of years, the author has taught a graduate-level course at Bradley University on the subject of management information systems. The course is aimed at teaching the basics of system analysis to people who will become the users of systems. It is based on the idea that the user is responsible for defining his or her own systems and that the user cannot delegate that responsibility to data processors. The idea of FA came from that course. It was defined in summary form in a 1987 textbook written by this author entitled *Managing Information Systems: An Integrated Approach*.

The technique of FA has been used on a number of classroom problems. It has also been used on a number of real-world business problems. It works! It works very well. For that reason, it seemed to be time to document the technique in some detail in the hope that others might find value in it.

The first, second, and sixteenth chapters of this book contain some material that appears in *Managing Information Systems*. The purpose of this repetition is not to add length to this book. The intent is to make the book stand alone as a complete reference for FA.

This text contains many examples of the use of FA in a manufacturing business. The concepts of JIT (just in time) and CIM (computer integrated manufacturing) have been included where appropriate. JIT and CIM are very much based on the use of computers to manage complex processes and large volumes of data. FA is based on the idea that a computer should be used only where it is absolutely needed. It is hoped that FA can help to manage the introduction of these new technologies where they can be of most value.

Thanks are due to more people than can be mentioned here. They were generous with their help and encouragement during the development and first use of FA. It takes a lot of patience to try out new ideas. They almost always sound idealistic—and often are.

Several people deserve special thanks for help in the development of this book. First is my wife Nancy, who is always encouraging. Professor Hirohide Hinamoto of the University of Illinois suggested the value of studying conflict within organizations. Professor K. S. Krishnamoorthi of Bradley University encouraged the use of functional analysis in a graduate course on managing information systems. Professor Alka Harriger of Purdue used a draft of this text successfully in a systems class. Sharon Maule of Caterpillar reviewed the first full draft in great and helpful detail.

I hope you find FA to be very useful and beneficial.

1
Introduction

INTRODUCTION

This is a book about the process of improving the procedures by which the day-to-day operation of a business is conducted. The procedures, the rules for how the business is operated, are the true systems of a business.

Today it is common to hear the words *computer system* used as a pair, like *damn yankee,* as if they are inseparable. The terms *computer* and *system* convey two different ideas and should be understood as such. If they are not understood as two separate ideas, then all our business systems will end up on the computer, not necessarily to the benefit of industry. The technique described in this book is aimed at the simplification and integration of information flow within a business. It is of secondary importance which systems end up on the computer, a subject dealt with in Chapter 16.

This is a book about information theory. Information is the coordinating medium that keeps the various units of the business operating in harmony. There are really three different types of things that flow through a business to keep it operating: information, materials, and services. The management of the flow of materials is a topic of much discussion today. JIT (just-in-time) management of materials is the "in thing." This book will not add much to the concept of JIT, but it will offer a way of dealing with the information needs of JIT. Services such as heat, light, ventilation, and cleaning are not within the scope of this book. But the book will offer a way of understanding how information is used to plan, order, and manage the services needed by a business.

This is *not* a book about computer theory, but the ideas it contains are vitally important to both computer people and line management. The information systems of a business are the responsibility of those who manage the business. Since the advent of the computer, there has been a tendency for managers to forget their responsibility for the management of information systems. There is also a tendency to forget that many information systems are not for the private use of each unit of the business. They are for the benefit of the whole enterprise. For this reason, information systems must be designed for, and by, the units of the business that they affect.

JAPANESE MANUFACTURING SUCCESS

There have been many theories that try to account for the loss of U.S. leadership in the production of commodities such as video equipment, cameras, and microwaves. Even major commodities are not immune to foreign pressures. The production of steel and the building of ships have moved to the Orient. The production of automobiles is in imminent danger of moving in the same direction. Many visits have been made to Japan, Korea, and Taiwan by political and industrial leaders. Much has been said and written about the probable causes of this manufacturing success in the Orient.

Theories are about as plentiful as theorists. A partial list of theories includes things such as:

- Sponsorship of industrial growth by the Japanese government
- Lower wages for workers
- A cultural tradition of a strong work ethic
- High levels of automation
- Unfair tariffs and exchange rates

Since there are so many theories, there might as well be one more on the list:

- Simplified, integrated systems

The typical large company in Japan has a practice of moving new management employees from department to department for the first five to ten years of their careers. This certainly gives them a broad exposure to the various aspects of the operation of the business. In addition, the Japanese are well recognized as being team players. They take great pains to seek agreement among all affected parties when a major change in operating procedure is being contemplated. They discuss the change at length and strive for acceptance by consensus. They adhere to the plan rigorously when it is finally put in place.

These practices are clues to what appears to be a Japanese determination to *refine the systems by which the business is managed.* Japan is moving to JIT manufacture and to JIT delivery of materials to the assembly area. This is a total rethinking of the production system.

Japanese manufacturers are marketing automobiles with minimal buyer options and many built-in luxury features. This is also the result of rethinking how the business should operate. It is a fundamental change in business strategy and has caused changes in many functional strategies and practices in the areas of engineering, manufacturing, and marketing. The whole operation of the business has been adapted to support a single corporate strategy.

PURPOSE OF THIS BOOK

The purpose of this book is to offer an alternative to the Japanese approach to systems. The intent is to encourage an integrated method of systems analysis.

All too often the different units within a business work independently to develop better systems. The systems are aimed at serving the best interests of the unit and rarely consider the best interests of the business as a whole. The result is frequent conflict among systems and between people. How can we compete successfully with the Japanese or with our domestic competitors when we spend so much of our energy competing within the business enterprise itself?

The method of systems analysis that this book offers is called functional analysis (FA) and is intended to aid in the development of integrated systems. These are systems that make it possible for the flow of information within the business to fulfill the needs of the individual units while enhancing the operation of the business as a whole. FA is aimed at helping each business unit to work in harmony with neighboring business units.

Good systems are those that pay heed to the things that motivate people. People do not provide information to their peers simply because they are responsible for doing so. The quality of the information provided by one person to others is often determined by how much trouble it is or what the reward is for doing so. Or the quality of the information is determined by what the penalty is for failing to do so.

It is not unusual for a person to pay little attention to what the customer would like in the way of information. This is especially true when the customer is a part of the business, a captive customer. FA offers ways of balancing the needs of the user of information with the motivations of those who are the source of the information.

BUSINESS SYSTEMS PLANNING TECHNIQUES

An approach called Business Systems Planning (BSP) has been used by industry for a number of years. It has been used very successfully.

A newer, Critical Success Factor (CSF) approach has become popular in recent years. It brings top-management attention to the importance of information systems to the success of the business as a whole. Critical success factors are those things that must "go right" in order that business goals might be reached.

Both techniques begin with business missions and strategies. So does FA. But the other techniques make too large a leap from strategies to systems and from systems to computers. In fact, there seems to be a predisposition in these techniques that every information need must be solved by computer. There seems to be little recognition that information problems are based in procedural problems. Unless the procedural problems are solved, use of a computer only covers up the problems. And once a procedure is mechanized, it becomes very difficult to change and improve it.

The various functions of a business are not independent. They must work together, not separately or in competition with each other. FA offers a way of making this happen.

FUNCTIONAL ANALYSIS

FA is a formal, step-by-step approach to the review of business practices. It leads to the restructuring of the flow of information within the enterprise for the welfare of the enterprise as a whole. FA offers a means for cooperative discussion and problem solving among the various parts of the organization. It makes possible the resolution of information problems for the best interests of the business and not for the private benefit of units within the business.

FA consists of a series of modest, logical steps. These steps lead from corporate strategies to operating procedures, to information systems, and even—where appropriate—to use of the computer. The series of logical steps for conducting FA are discussed in more detail in the next chapter. Chapters 3 to 16 of this book are each based on one of the steps and are cookbooks on how to be most effective in performing the analysis.

The theory in each chapter is supplemented with checklists that you can use to guide your effort. The checklists, as in Figure 1–1, are surrounded by a single boundary line and have no title above the text. The text also contains examples of the results of an exercise in FA, as in Figure 1–2. The examples are also surrounded by a single boundary line but have a title above the text.

There are 11 different types of documents that are produced during FA. Some of them are developed in stages through several steps of the process. A final example of each is shown in the Appendix.

DO IT YOURSELF

The technique of FA works for a subset of a business as well as for a business as a whole. This is an advantage over some other analysis techniques that must start the process at the executive office.

You, the reader, can profit from the use of FA in your own area of responsibility. The basics of how to do the analysis are very simple. Ask yourself what your product is (or what the products of your department are) and what your mission is. In simplest terms, your mission should be the delivery of your product to the customer. Ask who your customer is. Ask if your customer values your product and if he or she agrees with you on how to measure how well you perform your mission.

Do you have any conflicts with your customers, with those who

1. Read the book
2. Select a project
3. Perform analysis

Figure 1–1. Example of a Checklist.

DOCUMENTATION

The body of this form contains an example of the results of a step of Functional Analysis. The full set of this documentation is of permanent value as a set of procedures describing how the business operates. It also serves as a basis for change when the environment changes or when it becomes necessary to change the mission of the business.

Figure 1–2. Example of documented results.

supply products to you, or even among the subdivisions of your own organization? How might you resolve the conflicts?

Why not start small and evaluate the territory you know best? Solve your own problems before you worry too much about your neighbor's problems.

This book gives you a simple recipe to follow, chapter by chapter, step by step.

PRACTICAL EXPERIENCE

The author used FA a few years ago to analyze the flow of work in a business unit for which he had just taken responsibility. The unit consisted of five subunits and was responsible to develop and install a set of computer software for customers. The managers of the five subordinate units were taught the basics of FA and worked as a team to evaluate the flow of work and the flow of information through the organization. The results were impressive. One result was the elimination of some activities that proved to be of no value. Another result was the elimination of a number of points of chronic conflict among the managers. There was a general improvement in the effectiveness of the organization.

The second industrial use of FA was a study of three subunits working in the broad area of office automation. The subunits worked as if they were totally independent of one another. The result of the study was a rearrangement of work among the units and a new recognition of the degree of interdependence among them. Redundant effort was eliminated.

MANAGING THROUGH INFORMATION

People are not always aware of the impact of their work on their neighbors. They try to optimize their internal operation, often without thought about those whom they serve. It is often easier to work that way, but it is certainly less effective. Such a method of operation results in a poor supply of information for others. It leads to poor downstream operation, and it can make the downstream units look to other sources for the information that they need: "If you cannot give me what I need, I will do it myself."

All units of a business are in one way or another linked together by information. The individual units are not independent, even though they often act as if they are. If a unit does not perform properly, the whole business pays the price. Assembly of a quality product cannot take place without good material. A production schedule cannot supply the needed material without the information from a good sales forecast.

Information is the lifeblood of a business. The use of FA is a way to make sure that the arteries are free of needless blockage.

CAVEATS

The technique of FA is sound and proven, but it is not a guarantee of success. The recipe for applying the technique is very simple, but it is not easy to apply. It takes a lot of cooperation from people who are not in the habit of thinking in terms of information flow and processing rules.

FA does not yield instant results, but it does provide sound, lasting results. Time has shown that continued success cannot be taken for granted after use of FA. Any new systems will need ongoing measurement and control, the same as any other business process. For FA to succeed, it needs open-minded, cooperative people on the analysis team. This is the factor that makes the job difficult: People are not used to thinking about the best interests of *other* people.

There is a need for iteration when using FA. In step 5 of the process, you might learn something that can cause you to redo the work from steps 2, 3, and 4. This does not take a lot of time but does cause a lot of rework of written results. Word processing might well have been invented to serve the needs of FA.

It is recommended that you accept the definitions in this book. It would be wasteful of your time and temper to debate the meanings of the words *strategy* and *tactics, objectives* and *goals*. The definitions in this book work quite well within the context of FA.

SUMMARY

The whole subject and purpose of this book is functional analysis (FA). It is a sound technique to help you analyze and improve the systems of the business and the flow of information within the enterprise.

If you are not in a position to be interested in that subject, stop reading right here. If you are interested in making a major improvement in the operation of your business, read on—and good luck.

2

Functional Analysis

INTRODUCTION

The reason for doing a functional analysis (FA) is to make an improvement in the day-to-day operation of a business. It is a way of rethinking the rules by which the business runs.

The rules of operation are the heart of the system by which the organization runs. A system is the way in which people pass information to one another. The rules by which it operates determine the timing and quality of that information. Some of the rules are built into the computer systems of the business, but they are also built into the daily procedures for manual work. These rules might or might not be documented. Usually they are not.

It is not wise to assume that it always requires mechanization to make an improvement in any system. Many systems can be made to run better by means of a simple change in the rules by which people perform manual tasks.

FA is a series of simple, formal steps for examining system rules to see where change is appropriate. The structure of the technique is laid out in this chapter.

DESCRIPTION OF THE TECHNIQUE

FA begins with a statement of the business mission and the key strategies that direct the business. It then proceeds in modest, logical steps of in-depth analysis. It ends with a test to decide which of the systems of the business merit the use of a computer.

FA takes into account the fact that each part of a business is dependent on other parts. Information flows from unit to unit. The needs of the receiving unit are taken into consideration when defining the rules of operation of the supplying unit. Because of this concern for the flow of information from unit to unit, there is a need for teamwork among the people in the involved units. The analysis is not a "popularity contest." The purpose of the exercise is to define a set of rules of operation and a list of potential computer applications.

STARTING FOCUS

Is your business a new one? If so, then FA is a way of setting up your work rules in advance. Just as it is necessary to lay out the flow of material in a factory, it is necessary to lay out the flow of information in a business.

Are you a new boss in an old business? Then the use of FA can help you to understand what you have inherited. In addition, it is a way of giving the next level of managers a way of rethinking how they operate. This can be done without pressure to change methods just because a new boss wants to do things his or her own way. The author did just this thing a few years ago. The next level of managers accepted the use of FA as a way to tell the boss how they did things. Each also saw in the study an opportunity for making some other manager change what seemed to him or her to be a poor practice. And in the process of FA, each found reason to make change in his or her own area.

Do you have a tough problem to solve? FA can be a way to deal with it. FA is a way to obtain the involvement of those who are key to the solution.

Is your business one that has become "content"? "We do things this way because we have always done things this way." Then you surely need a way of waking people up to the thought that times have changed and that there might be a need to change the business.

Do you have a desire for improvement but do not know just where or how to make the needed changes? Then FA is a way to identify the opportunities.

DEPARTMENTAL VIEW VERSUS BUSINESS VIEW

Each department, each unit of the business, tends to look first at what is good for itself. Departmental budgets help to foster this sort of provincialism. If you do not have enough money to do what you think is necessary for other units, you tend to take care of your own department first. The other individual is left to his or her own devices to obtain needed information and services. This happens all the time. It can cause units of the business to take on added work to fill a void left by the failure of other units to provide what they need. If you go into the analysis of a large segment of the business, you are apt to find a need to change the organization as well as the systems within it.

FA has its greatest impact when used on a whole business. But it can also bring great improvement to a portion of a business. It is just a matter of defining the limits of the area to be analyzed.

To learn how to use FA, it might be best to start with a local problem. That way you can gain some experience before you start on the analysis of the whole business.

PRODUCTS AND MISSIONS

FA is based on a very simple model. Each unit of a business produces a product. If it does not, then there is no valid reason for the unit to exist. The mission of a unit is to produce a product for some other unit or units.

The product of a unit is one of three things. It can be *material,* as in the output of a factory. It can be *information,* as in the output of a marketing department. It can be a *service,* as in the output of a maintenance shop (see Figure 2–1). Of course, a unit of business

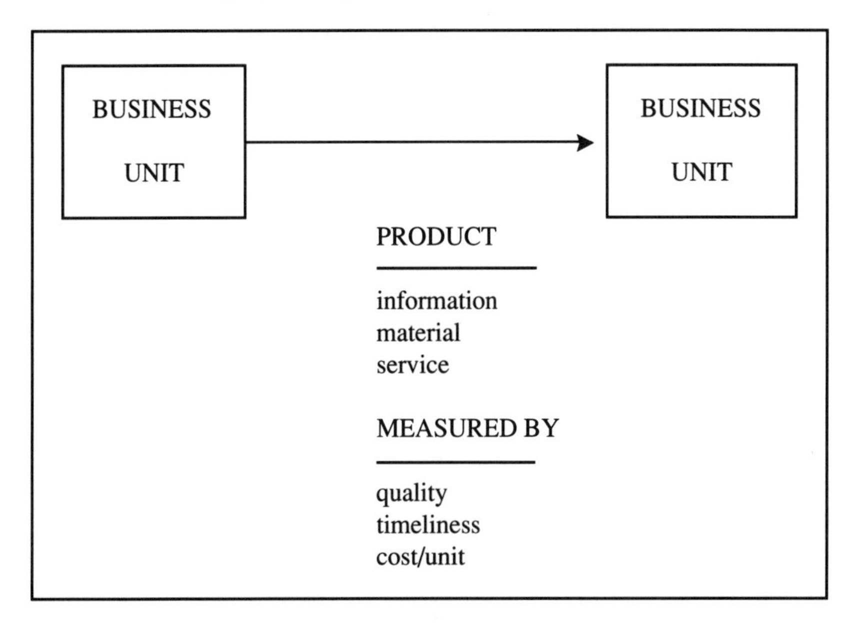

Figure 2–1. Function of a business unit.

can have more than one product. The factory no doubt reports the amount of production and scrap. But there always seems to be one product that is the main responsibility of a department—the primary reason for the existence of the department.

Customers come in two flavors: They can be internal or captive customers, part of the business itself; or they can be external customers who are totally free to buy or to reject the product of the business.

There are three measures of how well the supplying unit provides the product for the best interests of the customer. These measures are quality, timeliness, and cost per unit. The degree to which the department performs its mission is determined by measuring the product. There are three independent measures that are very useful and seem to be comprehensive. The first is the quality of the product, the degree to which it serves the needs of the customer. The second is the timeliness with which the product is produced, the degree to which it is available when needed by the customer. The third is the cost per unit of product. Refer again to the model in Figure 2–1.

The first two measures are best applied by the user of the product. After all, if there is no user, there is no reason for the product—or for the unit that produces it. The third measure, cost per unit, can be applied by the user or by the producer. If the cost is borne by the user, as in the price he or she pays, then it is a valid measure of the value of the product as opposed to other alternative products. If the cost is borne by the producer—as it most often is for products used internally in a business—then there is an immediate opportunity for organizational conflict. A tight budget has a way of making the producer reduce the quality or timeliness of a product or even cease to produce it. Obviously, this is not in the best interests of the user.

FA is a way of analyzing the flow of products within a business, with special emphasis on the flow of information. FA pays special attention to measures of the adequacy of the information as supplied from unit to unit.

SYSTEMS ARE NOT COMPUTERS OR SOFTWARE

A computer is a machine for processing data. Software is a set of instructions, a recipe, that tells the computer how to do its job. A system is a way of doing business. A system is represented in the rules, the conventions, the procedures that guide the daily operation of a business.

Computers and software can be used to make systems more cost-effective. But they are only tools to help the business operate a system. It is not good for a manager to begin to think that the systems of the business are a set of software. If he or she does so, there is the danger that the manager might lose a sense of responsibility for the systems.

One of the reasons that FA was developed was to help users retain control over their systems. The technique offers a nontechnical way of analyzing systems needs. It gives the user a tool with which to plan his or her own method of operation. It leaves concerns for computer technology to the computer people. "Render unto Caesar . . ."

SIMPLIFY BEFORE AUTOMATING

Simplify before Automating is the subtitle of this book. It makes a very important point: If you automate outdated and complex procedures, you make them even harder to change in the future. You freeze them into a permanence that is very hard to modify with changes in the business needs. But if you simplify operating practices before automating them, you make the original automation much simpler to accomplish. And you make it easier to modify the automated practices with changes in business needs.

FA is a very good tool to apply to your business systems in order to simplify them. But before you apply FA to old practices, look at the new computer technologies that are available. Ask yourself if any of them hold promise to permit you to make fundamental changes, improvements, in operating strategies. An example might be the possible use of an automated storage/retrieval system to reduce the cost of storing and accessing production materials.

SEQUENCE OF STEPS

As stated before, FA proceeds in a logical series of steps, as shown in Figure 2–2. The process begins with a description of the business, its mission, and the metrics by which the performance of the business is measured. The key strategies of the business and its departments are documented in steps 5 and 9. These are an outline of the

1. Business description
2. Business mission
3. Measures of business performance
4. Business goals
5. Business strategies
6. Department mission
7. Measures of department performance
8. Department subfunctions
9. Department strategies
10. Organizational conflicts
11. Information flowchart
12. Subfunction description
13. Rules to resolve conflicts
14. Computer systems requirements

Figure 2–2. Functional analysis steps.

systems of the business. The systems of the business are spelled out in some detail in steps 11, 12, and 13. In the last step, number 14, a test is made to decide which of these systems merits mechanization.

Some of the steps of FA can be done in parallel, as shown in Figure 2–3. There is little problem with performing some of the steps at the same time, since not all the logic is linear. But as business strategies are verbalized, there is the possibility of finding the need to redo some departmental missions and measures. If time allows, it is best to follow the linear sequence.

It is well to recognize that there will be a need to recycle thoughts,

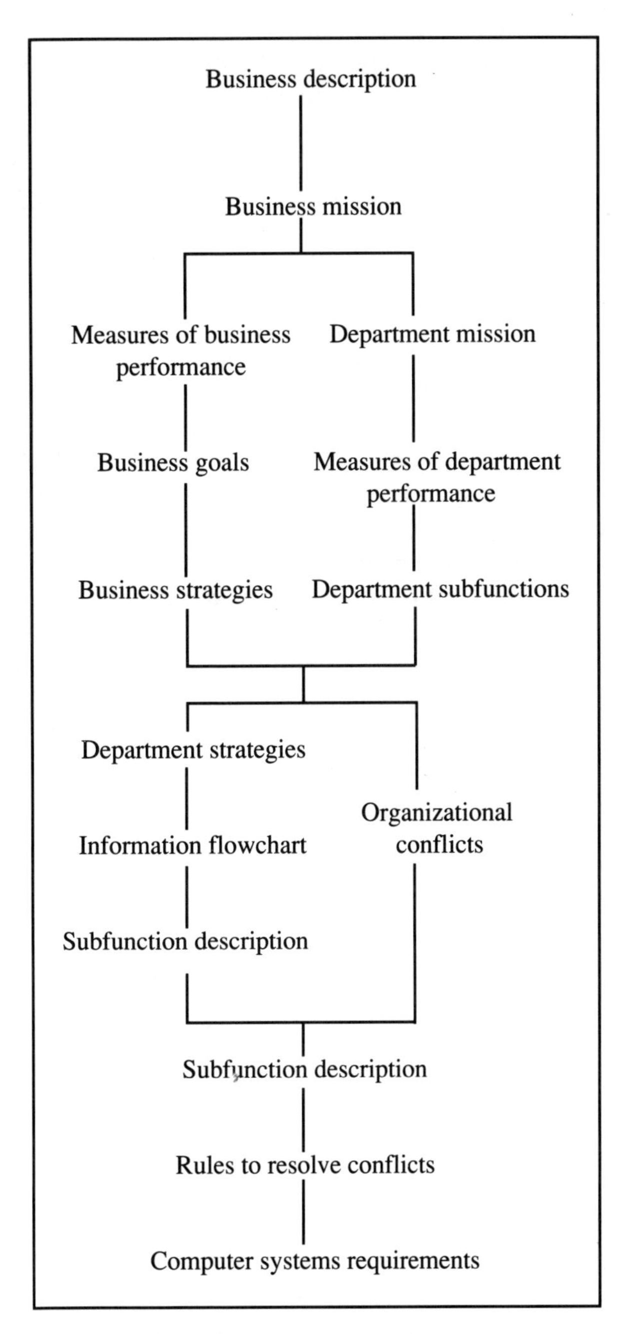

Figure 2–3. Sequence of steps.

even if you follow the full sequence of steps. Each step of thinking deeper into the operation of the business is a step toward discovery of new needs and new solutions. The process of systems analysis is like removing the rocks from the bed of a stream: Until you have lifted a big rock, you have no way of knowing how many smaller rocks lie hidden beneath it.

GOALS AND OBJECTIVES

It is wise to distinguish between goals and objectives when doing FA. Goals are specific measures of the level of performance of the day-to-day business. An example would be a scrap rate from production that is not to exceed one piece in a thousand. An objective is like a goal but with a specific target date. It is something that is to be obtained after a specific change is made. An example would be a scrap rate of less than five pieces in a thousand by the end of June. It is based on the assumption that the means are not yet in place to attain the desired level of performance.

These definitions might not be natural to you, but the process of FA will go better if you accept them. At least accept them for the duration of the study.

ORGANIZATIONAL IMPACT

An open-minded use of FA can well result in the need to change some part of the organization of the business. The structure of the organization itself can be one of the primary causes of conflict. The higher one has to go in an organization to resolve a difference of opinion, the harder it is to obtain a resolution. This can cause managers to avoid identifying conflict, since it is easier to pretend that none exists. The same is true if it takes the involvement of too many people, at any level, to resolve a conflict or to reach a decision.

PLANNING TEAM

The more complex a systems project, the less likely it is that a single person in one department can make good decisions for people in other departments. In fact, the less likely it is that the people in other departments will accept the decisions, good or bad. The list in

1. Decision-making body
2. Thoughts of all functions
3. Nobody knows everything
4. Cooperation

Figure 2–4. Reasons for a systems team.

Figure 2–4 shows the major reasons why a systems team is needed in order to perform FA. Departments are likely to be reluctant to delegate authority for decisions affecting them. So it is necessary to establish some sort of decision-making body with representation from each of the involved units. All the decisions will not be popular in all the units, although each unit will have to recognize that it has had its day in court. Many times a decision that is unfavorable to one unit has a very significant advantage to other units and to the business as a whole.

The information and material flow through a business can be difficult to map and to understand. It is not an easy task to evaluate the pros and cons of new ideas of how to conduct the business. For these reasons, it is desirable to have a forum in which the thoughts of each functional area can be represented. Nobody knows everything, especially the person who seems to think that he or she *does* know everything. A team of people from varied backgrounds is a big help in supplementing the experience and judgment of the system designer.

No matter how well a system is defined, designed, and constructed, it will not operate well unless the people who have to work with it have "bought in" on the idea. The best way for people to take on a feeling of responsibility for a new system is to have them deeply involved in the definition and in the installation planning. The Japanese are well aware of this. They go to great lengths to allow affected people to discuss a new system before attempting to implement it. But once the discussions have been completed, woe to the person who shows a negative attitude toward the project.

CHARTER

An FA study should begin with a charter. The authority of the team is best established by a written statement from the high-level sponsor of the project. Otherwise, there is a danger of a lack of cooperation, and even overt hostility, from people within the organization.

The charter should set out the limits of the study. It should also describe the general scope and the expected results of the study. The scope could be as broad as to refine the information flow within the whole company. The scope could be as small as to refine the method of reporting the effective dates of product changes. The charter should also set an expected target date for the completion of the study in order to retain a sense of urgency. If the charter names the team leader, it gives him or her added personal authority. And the establishment of a target date helps the team leader to gain the commitment of team member time for a protracted period.

TEAM COMPOSITION—FUNCTION

The major responsibilities of a systems team are listed in Figure 2–5. The total list for any specific project could be much more detailed, depending on the size and nature of the project and the culture of the company.

A most important part of the FA process is the generation and resolution of ideas for procedural rules. Remember that procedures are the true systems of the business, and the impact of new procedures can be far-ranging and subtle. It makes a big difference if materials are ordered daily, weekly, monthly, or on demand. It makes a big difference if an assembly line runs one, two, or three shifts per day. It makes a big difference which department owns and is answerable for the inventory of finished product. It could be the Assembly, Master Schedule, Marketing, or Sales Department. Each would tend to manage the inventory by different rules in order to serve its own needs as well as the needs of the business.

The motivation for development of a new information system is often the hope of resolving some major business problem. The solution to the problem is usually much less apparent than the problem is. A team effort is a good way to generate a set of alternative solutions and to select among them. This was well demonstrated during World War II when the British used teams of various skills to solve strategic war problems.

1. Resolution of procedural rules
2. Generation of solutions
3. Evaluation of mutual impact
4. Significant decisions
5. Resolution of conflict
6. Project justification
7. Progress monitoring
8. Installation planning
9. Communication and buy-in

Figure 2–5. Systems team functions.

The team must evaluate the impact of the proposed new rules on each department. If Assembly operates on a three-shift basis, then how will Materials Distribution stock the material storage bins in the assembly area?

The operation of a major portion of the business is much more complex than it appears on the surface. There is nobody who is so brilliant and incisive that he or she can propose radically new procedures without the need to check out the implications throughout the organization. A team is needed for this. When a significant new procedure has been defined, it usually turns out that there is no single authority who has the insight to approve or disapprove it. It is probably too complex an issue to be explained to the chief executive officer. In effect, the team itself becomes the decision-making authority for most major changes, because there is no other knowledgeable authority. The team has the authority of knowledge.

In the resolution of business problems, there is bound to be a great deal of organizational conflict. For almost any significant issue, one portion of the organization must make concessions for the benefit of others. These concessions do not come easily, especially if budgets are tight. Often the resolution of a conflict is not accomplished through the cold application of business logic but through peer pressure.

When a new system is put into place, it usually causes a significant change in the way different parts of the business do their work. It adds effort and cost to some, and it reduces effort and cost for others. Input is needed from all disciplines to evaluate the expected costs and benefits from the proposed changes.

As work progresses on the FA project, a team is needed to review progress. It can make sure that no significant portion of the work is falling behind to the detriment of the schedule for the whole project. A team can evaluate alternatives when a portion of critically needed work falls behind or proves to be infeasible. Of course, group pressure helps to motivate units of the organization to meet the agreed-on schedule.

When the FA study is complete, the team might have a new task. A new series of steps must be completed when the study shows the need for mechanization. It takes mutual planning and agreements to identify and to schedule the various tasks that must be done to develop and install the application. These tasks can include such things as data conversion, user training, system testing, and checkout. The conversion of data usually results in a great amount of "grief." *Grief* is a term that is used to describe old data records of demonstrated inaccuracy or data records that are not complete. The team can set targets and plans for resolution of the grief. The detail tasks of this follow-up activity to FA are not the subject of this book. They are well described in other texts.

As discussed before, it is vital for the affected units of the organization to "buy in" to the concepts of the new system and to the expected changes in operation. Team members from the various departments are the first people to make that buy-in. They do this by means of the agreements they reach in team meetings. They then have the responsibility to take the agreements back to their respective units for discussion and for acceptance by their peers.

TEAM COMPOSITION—RANK

The departments that must be represented on a systems team are simply those that are most likely to be affected by the proposed system. The team should also include representation from any department that can contribute to the study. For example, Accounting and Data Processing often have value to offer to any systems project. The primary qualifications of team members are listed in Figure 2–6.

The team should consist of broadly experienced people who can generate new ideas and who have the background to evaluate the possible impact of new procedures. The greater the breadth of experience of the team members, the more likely they are to be open-minded about proposals that put new demands on their own departments. This is one area in which the Japanese have a great advantage, since they make a specific effort to expose new management employees to a number of different functional areas. The typical Japanese manager has acquired a broad knowledge of how the business operates outside his or her own area of responsibility.

The team should include a representative of the Data Processing Department who could be the primary architect of new mechanized procedures. This person is the one most likely to be skilled in systems analysis and documentation techniques. The Data Processing representative is also the one who is most likely to have a broad insight into the inner workings of a number of functional areas. He or she has probably worked on development of mechanized systems for many departments.

1. All involved departments
2. Broadly experienced people
3. Data Processing representative
4. Middle-level managers
   - know the system
   - make decisions
   - able to compromise
   - influence within department

Figure 2–6. Systems team membership.

When a unit has to assign a person to an extra, time-consuming task, there is a temptation to assign the person who can best be spared. This is usually the wrong person. It is in the unit's best interest to assign a top-level thinker. The head of the unit is usually not the right person, since he or she often has only a surface-level knowledge of the systems by which the organization runs. Budget and personnel requirements tend to occupy much of the department head's time. This prevents him or her from staying current with procedural detail. A clerical person or a first-line supervisor is also probably the wrong person to assign to the team. The lower the level of the person in the department, the less likely that the person has a good perspective on departmental procedures and their impact on neighboring departments.

A middle-level manager or a high-level staff person usually has a good perspective on the systems of the department and on relationships with neighboring units. Such a person is also likely to have the perspective and the confidence to make decisions that will affect the department. He or she is also likely to have the needed respect within the unit to obtain concurrence with the decisions.

EXAMPLES OF TEAMS

If the topic of the study is a problem in Inventory Control, the team should include middle-level managers from that department and from related departments. These might be Purchasing, Manufacturing, Assembly, Shipping, Product Support, Quality, Accounting, and Data Processing. Each has an impact on the level of inventory, is affected by it in some way, or helps to measure it.

If the problem has something to do with the line of product that the company offers, the team will be different. It might include Design Engineering, Marketing, Sales, Pricing, Product Support, Accounting, and Data Processing.

TEAM LEADERSHIP

Figure 2–7 lists a number of the key qualifications for a team leader. The primary requirement of a team leader is that he or she be respected and accepted as leader by the members of the team. Without this, any other qualifications are wasted. The leader should be reasonably highly placed in the organization, usually at least as

1. Respected by team
2. Highly placed in the organization
3. Knowledgeable of the territory
4. Good systems comprehension
5. Good leadership skills
6. No "axe to grind"

Figure 2–7. Systems team leader qualities.

high as the highest ranking team members. Rank brings a degree of respect that sometimes is not gained by other qualifications. But rank alone will not retain respect if the designated leader proves to be inadequate to the task.

It would help if the leader were broadly knowledgeable of the territory to be covered by the envisioned system. This knowledge will stand him or her in good stead when there are strong differences of opinion among the team members. Most often, team members are better at finding deficiencies in one another's ideas than they are in generating new ideas themselves. It takes a strong leader to keep the flow of thought moving in a positive direction.

The leader will have a distinct advantage if he or she has good systems sense. Some people seem almost intuitively to be able to track cause and effect in the flow of information and materials through an organization. Data processing experience and participation in systems work are the usual ways by which a person learns to detect and solve systems problems.

Of course, the leader should have recognized leadership skills. He or she should be able to plan meetings and keep the discussions on track. He or she should be able to delegate responsibility and hold subcommittees to the completion of their assigned tasks. He or she should be able to keep discussions active without losing input from passive team members and without allowing the domineering ones to force half-formed ideas on the team. Team meetings are a forum for analytical thinking, not for selling political slogans.

One of the most important characteristics of a good leader is that he or she does not appear to have a preconceived bias about the area under discussion. If the leader appears to have a private "axe to grind," he or she will find the members to be polarized and unwilling to seek the necessary compromises.

TIME INVESTMENT

Do not hurry the process. FA works best if people take the time to think about each step. It takes time to identify problems. It takes more time to accept them as being real and in need of solution. Then it takes more time to develop solutions, and it takes yet more time to accept the solutions and to implement them.

The frequency of team meetings is to a great extent a function of the culture of the company. Meetings that occur less often than once a week are a sure way of losing continuity. You spend all your time catching up with where you are. At times, it is of help to set one or two days aside in order to think through a stage of the process thoroughly. An example might be stage 5. The individual team members could have been charged with the task of writing down their ideas of business strategies. These, by the way, are usually undocumented. Then the team as a group could discuss the strategies and try to select the most meaningful ones. This takes dedicated time. It is hard to do this in a short meeting. After a long working session, it is best to let people meditate on the results for a week before meeting again to resolve any second thoughts.

Meetings for the purpose of developing plans and work assignments should only last an hour or two. This is also the best rule for meetings to review the results of workshops. A good rule of thumb is for normal meetings of two hours or less and workshops of one or two days.

DOCUMENTATION OF RESULTS

Each meeting should have a note taker. It is best if this person is not one of the team members. You want the team members to be able to devote their full attention to the evaluation of ideas. The results of each meeting should be distributed to the team members in time to be digested before the next meeting.

The documentation of the results of FA should not be done in great detail. If it is, it will never be read or evaluated—or acted on. Examples of FA documentation are shown in succeeding chapters. They are designed to be terse without losing meaning. Some of the text in this book is lengthy in order to tell you how you can avoid being long-winded in documenting the results of your work.

It is best to use a word processor to record the results of the study. There are a number of software packages for drawing flowcharts to help you to document the work of stage 11, described in Chapter 13. Mechanical documentation will ease the recycling of ideas as you find reason to improve on them.

Do not allow yourself to be forced to handle issues out of sequence. Often there is somebody who has a "hot item" that must be dealt with in order to keep the peace. Write down the topic and agree on the stage of the process at which you will be prepared to deal with it. It is rarely necessary to resolve the issue immediately. It is always prudent to give the issue the respect of recognizing it as being important. To fail to do so is to take the chance that the author of the idea will be alienated from the process.

Every successful project has a high-level sponsor —somebody who ranks higher than any team member and who is ready to act on the results of the team. You need to make regular reports to this sponsor. This can best be done by showing this person the documented results of each stage of the study. It will make him or her more ready to accept conclusions that might otherwise be traumatic. This is a way of helping him to buy in to the results.

INVOLVEMENT OF STAFF

Although staff members are not usually on the project team, they have a valuable function to perform. The results of each stage should be reviewed with staff in each department. This is a way of finding errors in the team's thinking. It is also a way of getting buy-in from those people who will have to work with the new systems.

SUMMARY

FA is a logical, step-by-step process to rethink and to refine the systems of an organization. FA is not an exercise in automation for the sake of automation. The subject of automation should usually be left to the end of the project. The only exception is where technology is identified as a necessary means to implement a strategy. An example of this is an automated storage/retrieval system to support JIT production.

The careful use of FA should result in enduring improvement in systems. But it will take ongoing effort to implement the results of the study. It will also take ongoing efforts to track progress and to make change where change becomes needed. After all, the world is changing even while we try to decide on how to adapt to past changes.

An FA study should begin with a charter, which is a written statement by the sponsor of the study. It should clearly spell out the scope of the study, both in content and time span.

3
Business Description

INTRODUCTION

The best place to start a functional analysis (FA) is with a statement describing the business that is to be analyzed. If only a portion of the whole business is to be analyzed, it is still best to start with the documentation of the whole business. Then prepare a second statement describing the portion of the business in question.

This effort does not take long. It has the advantage of giving a focus and a boundary to the area of study.

DICTIONARY DEFINITION

A definition of the word *business* might aid understanding the material in this chapter. According to *Webster,* "a business is a commercial activity which is engaged in as a means of livelihood."

Since the subject of the chapter is "business description," a definition of the term *description* is also helpful. *Webster* says that a description "is an account of the properties or appearance of a thing, so that another person may form a concept of it."

WORKING DEFINITION

Now we can combine the two definitions into a single definition for you to work with in FA. For the purpose of FA, a *business description* is an account of the properties of a business and possibly of a subset of a business. It is used to give the team a common concept of the activities to be analyzed. This analysis is for the purpose of improving the operating systems of the business.

COMPONENTS OF A BUSINESS DESCRIPTION

The description of the business need not be a long document. In fact, it is best if it is a short one. It is hard to focus on a complex,

1. Description of the company
2. Description of the product
3. Description of the marketplace
4. Description of the physical facility
5. Major operating practices

Figure 3–1. Components of a business description.

wordy picture. The description should include the items shown in Figure 3–1.

The document should start with a brief description of the company itself. It should contain a description of the product or products of the company and should include a description of the marketplace, the set of customers for whom the product is intended. It is often convenient to describe the physical facilities in which the business operates. The limitations inherent in bricks and mortar have an impact on the degree of freedom to change operating practices.

Clues to aid the analysis can be obtained from identification of the major operating practices of the business. Do not try to be exhaustive, since most operating practices will be identified in detail later in the study.

EXAMPLES

Two examples are included to illustrate the scope of the description and to show a possible format for recording it. The first example is a fictitious business called Hydroclean, shown in Figure 3–2. Hydroclean is a manufacturer of washers and driers. Note that the full business description is no longer than a single sheet of paper. The second example is another fictitious business called Gardentool, shown in Figure 3–3. Gardentool is a wholesaler of lawn and garden products.

HYDROCLEAN

Description of company

- $200 million annual sales
- Manufacturer of washers and driers
- Located in the Midwest

Description of product

- Three models of each
- Four color schemes for each — baked on
- Retailer's label glued on when requested

Description of the marketplace

- Department store chains
- No direct retail sales

Physical facility

- Two assembly lines, each capable of handling both products
- Warehouse for storage of finished product
- Manufacturing area for frames, drums, and finish panels
- Subassembly area for control panels and heating units

Major operating practices

- 40-hour workweek
- Purchase of hardware, motors, and electronic controls
- Major change each model year
- Only safety or emergency change within a model year
- Part number change if a changed part is not interchangeable with the old part
- Product is built to stock and is sold from the warehouse

Figure 3–2. Hydroclean business description.

GARDENTOOL

Description of company

- $20 million annual sales
- Wholesaler of lawn and garden products
- Located in Atlanta

Description of product

- Tools and machinery from manufacturers
- Fertilizers and pesticides
- Seeds and bulbs packaged under own label

Description of the marketplace

- Department store chains
- Hardware stores and nurseries

Physical facility

- Display and sales center attached to a 20-foot-high warehouse space
- Rack storage of materials
- Packaging and labeling of seeds and bulbs in a workshop adjoining the warehouse

Major operating practices

- 40-hour work-week
- New catalog each year based upon new products and response to old products
- Prices maintained through the year with heavy discounting at end of season
- Delivery from manufacturer direct to customer for bulky machinery

Figure 3–3. Gardentool business description.

METHOD OF REPRESENTATION

The description should be written as brief phrases or bullets. The intent is to provide a snapshot that can serve as a reference document and that can be absorbed by a quick reading.The team needs something they can remember and refer to quickly as a focus for their discussions.

ANALYTICAL METHOD

Figure 3–4 summarizes a recipe for development of the business description. The recipe need not be followed rigorously. It is merely a guide to help you avoid errors or omissions.

A good source of information is any published prospectus on the company. It might be something you distribute to customers or to candidates for employment. The trick will be to find and summarize the key ideas, since the document will certainly have many times more words than you need. An alternative source of information could be recent annual reports to stockholders.

The team should develop the business description and agree on it in its final form. The content of the bullets can be selected in a brainstorming session. One or two people can write the document. Then the whole team can critique the resulting draft. It is often a waste of time if people bicker over words. But in this case, it might be well to allow some bickering to take place. It is a way of letting people buy in. By limiting the number of people who write the draft, the amount of needless argument over words will be minimized.

But having said that, do not spend too much time on this stage of the project. There is more than adequate time to revise the description because of what will be learned in later stages of the study.

1. Published prospectus

2. Team thought

3. Do not overdo it

Figure 3–4. Business description analytical method.

COMMON ERRORS

As said above, too much detail is bad. The team can be bogged down by concentrating on trivia. The leader's skills will be stretched from the first to keep the team making progress without alienating those who are detail-minded. The team members already know enough detail about the business. If they did not, they probably would not be on the team.

Try not to argue about what should be or what might be. That will come clear during the course of the study. It is the purpose of this stage of the project to document *what is*. Later steps of FA will be devoted to careful study of what should be or might be.

CAUSES AND CURES

The picky detail artist is a curse on this sort of a project. He or she is the one who wants to "word smith" everything and tends to argue that you cannot leave out anything that he or she thinks is important. If this individual is not constrained, the result will be a long document, a lot of wasted time, and a very frustrated team. It is best to screen team candidates. If one of this type gets on the team, then try to replace him or her, or have the individual keep private notes for future checking against the results of each stage of the study. But do not let him or her set the pace for the effort.

A second source of problems is confusion as to the mission of the team, which comes from a lack of clear understanding of the charter of the group. A good way of reaffirming the area of the business to be addressed is to review a draft of the business description with the sponsor of the study. The charter of the group should be in writing before even starting FA.

The business might be so varied that a clear description cannot be obtained—likely to be the case with a conglomerate of businesses. It is worth considering that the product of a conglomerate might be profit for the owners. Or you might subdivide the conglomerate into separate businesses and study only one of them at a time.

QUALITY CHECKLIST

Look at the checklist in Figure 3–5, which is designed to help you make a quick check of the quality of the document that you have produced at this stage. The document should be limited to a single page. Two pages are the maximum and should be avoided if at all

1. One page long, preferably
2. Two pages long at maximum
3. Each of the five headings included
4. No more than six items under a heading
5. Terse, descriptive phrases
6. Team general agreement as to content
7. Sponsor agreement as to content

Figure 3–5. Business description quality checklist.

possible. If you fail at this, then you have let the "nitpicker"have his or her way—and you will pay dearly for it in later stages of the study.

Does the document include all the major topics listed in Figure 3–1? Each is important. Is each topic covered by a small set of terse, verbal bullets? No full sentences are allowed. Limit the bullets to no more than six for each topic. Is there general agreement as to the content and wording of the document? If there is not, then it might be dangerous to continue. It is hard to obtain agreement at later stages if agreement has not been reached at the first stage.

Remember to review the document with the sponsor to obtain his or her agreement that it is accurate. If the sponser wants it changed, then change it. After all, the sponsor is the boss.

SUMMARY

The business description is the starting point of the FA study of a business or of a part of a business. It should be approved and accepted by the whole team as a basis for the study. It is best to start the project with a charter statement from the sponsor. It is also good to review the business description with the sponsor at the end of this phase. This should be used as an occasion to reaffirm the charter. Remember that the business description is subject to change based upon knowledge gained in later stages of FA.

4

Business Mission

INTRODUCTION

An important early step in Functional Analysis (FA) is to decide what business you are in. This is done by developing a brief mission statement for the business. Although the business description itself was brief, this more basic statement has distinct added value.

Most people have heard of the railroad company that decided it was in the transportation business rather than the railroad business. As the trucking industry became more competitive, the volume of freight shipped by rail began to decline. Many railroads ceased to be profitable, but this one railroad developed the concept of piggyback freight. It was the start of the shipment of semiloads all over the country. Obviously, a mission statement can be a powerful thing.

Although a mission statement is an image of the essence of the business that is in place, it is also an analytical step toward what the essence of the business might be in the future. It is an exercise in thinking about the business in a new and structured way. Certainly this was the case for the railroad. The business was the same at the moment the mission statement was developed, but its future potential changed in the same instant.

A portion of a business (a business within a business) might have little freedom to define its own mission. But it is still of great value to identify what that mission is.

The development of a mission statement is also a technique for thinking through the concepts of a new business. If you are investing in your future, it is well to have a clear statement of the direction in which you intend to go.

DICTIONARY DEFINITION

It is always prudent to start with a definition of the subject matter. This helps to avoid misunderstandings based upon assumed meanings of the terms used. *Webster* defines a *mission* with these words: "A mission is the sending of an individual or group by an authority to perform a specific service."

The broad use of the term *mission* by industry has led to the evolution of a new meaning: "the function for which a unit of business is responsible." This should not be confused with the military use of the term *mission*. In the military, a mission is usually a specific tactical objective. In FA the term *mission* describes a function, an ongoing activity that takes place from day to day.

WORKING DEFINITION

With these definitions as a base, the definition of *business mission* can be developed for the purposes of FA. It is the responsibility "to deliver a product to a customer." That product can be material, or it can be information or a service.

If a product cannot be identified, it suggests strongly that there is no mission and that the unit of business does not need to exist. The same is true about the need to identify a customer. If there is no customer, there is no need for the business. Do you remember the old story about the buggy whip factory? The mission of the company was to manufacture buggy whips. The company failed to recognize that its number of customers was declining as people bought autos to replace their horse and buggy mode of transportation. In time, the company failed. It still made a quality product, but it had no customers.

COMPONENTS OF A MISSION STATEMENT

The components of a statement of business mission are listed in Figure 4–1.

Do something. The business manufactures, or it sells, or it delivers, or in some other way, it makes the product available.

With a product. The product can be material, such as a washing machine. It can be information, such as a stock report. It can be a service, such as cleaning an office building.

(From a source). This portion of the mission statement is optional. As an example, a trucking firm delivers product from suppliers to users.

1. Do something
2. With a product
3. (From a source)
4. (Through an agent)
5. For a customer
6. (For a purpose)

Figure 4–1. Components of a mission statement.

(Through an agent). This is also optional. Some companies, for example, manufacture a product for sale through wholesale or retail houses. They do not deal directly with the end customer. This has a significant impact on how they advertise, sell, package, and ship their product.

For a customer. The customer can be an end user, or the customer can be an intermediary, such as a wholesale house.

(For a purpose). This option is less often needed than the others. The business can offer its product for a specific use by the end customer or by an agent. An example is shown in Figure 4–3, where the purpose is resale to the end users.

EXAMPLES

A mission statement for Hydroclean is shown in Figure 4–2. A mission statement for Gardentool is shown in Figure 4–3. All the examples in the rest of this book are derived from one or the other of these two fictional businesses.

METHOD OF REPRESENTATION

A simple statement of limited length is the best way to represent a mission statement. This is very difficult to do, since there is a natural desire to make the statement "complete." Most people, left to their own devices, would write a paragraph or a whole page of detail. This must be avoided. A simple statement is the best way to capture the essence of the business.

HYDROCLEAN BUSINESS MISSION

Mission of the business

To market company-built washers and driers to retail chains

Business parameters

- The product is aimed at home use, not commercial use
- The product is sold under the label of the retail chain if requested
- The product is not sold in countries other than the United States
- Customer service is provided by the outlet, with training by Hydroclean

Figure 4–2. Hydroclean business mission.

There is a need to allow for some way to add important details. The mission statement can be supplemented as desired by the addition of business parameters. These are statements that describe the scope of the business. They tend to represent strategies. They also show some of the differences between the business and its competitors. These added statements are called *business parameters* in the examples.

Some typical types of business parameters, listed in Figure 4–4, might describe:

- The type or variety of product
- The types of targeted customers or the distribution of customer population
- The method of production or the method of procurement of materials
- The method of delivery of the product to the customer
- The method of ongoing support of the product when it is in the hands of the customer

GARDENTOOL BUSINESS MISSION

Mission of the business

To wholesale lawn and garden products to retail outlets for sale to residential customers

Business parameters

- The product is aimed at home use, not commercial use
- The product is sold under the label of the manufacturer
- Only hardware items are sold outside the continental United States
- Customer service is provided by the outlet, with training by the original manufacturer
- Seeds and bulbs are packaged by Gardentool and sold under that label

Figure 4–3. Gardentool business mission.

ANALYTICAL METHOD

The typical steps for developing a mission statement are shown in Figure 4–5. Although the resulting statement is simple, the task of writing it is not. There is a well-known quotation from Abraham Lincoln in which he apologized for having written a long letter for lack of time to compose a short one. It is very important that these statements be terse. The same is true about all the documented results of FA. There is no value in volume alone. The imagery of the statements must be easily read, understood, and remembered in order to be useful.

The first step is to identify the current mission, which may or may not already be documented. There are two primary sources for this

1. The product
2. The customer population
3. The method of procurement
4. The method of production
5. The method of delivery
6. The method of product support

Figure 4–4. Business parameters.

information. The first is the same as for the business description. Look in business abstracts and annual reports. The second source is in the minds of the team members. They all know what sort of business they are in. "Everybody knows that." But if they are typical, they have rarely thought about writing it down.

The verb is the critical element of the mission statement. What does your company do? Does it produce, manufacture, sell, wholesale, design, advise, or distribute? Is the company driven by its production skills? Is it driven by a marketing man? In the mid-1980s, Firestone changed from being a manufacturer to being a sales organization because of a change in top management.

1. Identify current practice
2. Use action verb
3. List "commissioned" changes
4. Do not debate boundaries
5. Verify with sponsor

Figure 4–5. Business mission analytical method.

List any "commissioned"changes in current practice that were given as a part of the charter to perform the study. Yours might be a case like Firestone's. The sponsor of your team might have in mind a major change in emphasis on how the business operates. This is the place to put it down in writing, succinctly and clearly.

There will be other possible changes in the scope of the business. They will be developed as a part of the business strategies step in Chapter 7. Again, do not try to be exhaustive, since there is plenty of time to make changes and additions in later stages of the analysis.

Do not debate the business parameters in depth. Write down the thoughts that come readily to mind. Accept the fact that they can be clarified in later steps. Review the results of this step of FA with the sponsor of the project. The statement of business mission is a simple document, but its implications are profound.

COMMON ERRORS

The most common error comes from a natural desire for completeness and accuracy. The result, if this tendency is not controlled, is a long, rambling statement of mission. The long statement might make the authors feel good at first, but it will result in confusion. In order for a mission statement to be useful, it must be short and simple. In this case, clarity comes from brevity.

The most common error is the use of the word *and*. It makes it simpler to write, and it permits more to be said, and it helps to avoid omissions, and . . . Resist all temptation. The word *and* just adds length. If there is an overwhelming need to use the word, try adding the necessary information in the list of business parameters. That is the reason for adding them. They make it possible to keep the mission statement simple and easy to grasp.

Another common error is to fill the statement with adjectives and adverbs to describe a degree of performance or quality. Resist this. Performance and quality targets will be developed in later stages of the study. A statement like "to be the low-cost, high-quality source of world-class widgets"is better said as "to manufacture widgets." Excess verbiage just adds confusion.

Now that you have avoided the word *and* successfully, you might find that you need too many business parameters. You can use the word *and* here if you must, but you should only have five or six statements at a maximum.

Probably the most difficult constraint is the advice to avoid the word *and*. Note that the author used it in the example in Figure 4–2. The words *washers and driers* could have been replaced by the words *laundry equipment*, but that would not have been as precise. There can be exceptions to the rules. The best advice is to try to follow the recommendations of the text, supplemented with your own good judgment. But remain loyal to one primary rule—be terse.

CAUSES AND CURES

By now you should have a clear idea that the number of words should be kept to a minimum in the documentation of the results of your FA. This is true at all steps in the study.

One cause of excess words is an attempt to "word smith"for advertising effect. You are not writing for the public but for the business. You need to strive for brevity and clarity of image. Be ruthless in your effort to avoid adjectives, adverbs, and conjunctions. Do not include goals in the mission statement or even in the business parameters. "To become the low cost manufacturer of premium . . ." is at best a goal, and at worst advertising. If there is any confusion at this point about the meaning of the term *goal*, refer to the definitions in Chapter 6.

Another source of confusion and delay is to try to strategize or to make decisions too early in the process. Take notes of strategic statements as they come up in discussion. Save these for a later step in Chapter 7. This is a good general rule. If there is any doubt about detail, note it down and set it aside for future reference. Keep the published statements simple.

Some people attempt to lobby for elusive future goals or strategies too early in the process. Again, keep note of these items and wait one or two steps. The process is iterative. If anything important is missing, you can always add it to the mission statement later—one of the benefits of using a word processor.

The mission document can become too lengthy if the team attempts to be all-inclusive. The leader must encourage the team to seek clarity, not completeness. Keep it short and simple.

QUALITY CHECKLIST

Refer to Figure 4–6 for a quality checklist for the mission statement. It will help to remind the team to "keep it simple."

1. Begins with the word *to*
2. Less than 15 words long
3. No "advertising" adjectives
4. No "advertising" adverbs
5. No quality descriptors
6. No more than a page long

Figure 4–6. Business mission quality checklist.

The mission statement always begins with the word *to* and should be limited to less than 15 words. This is not an absolute rule. But if the number of words is allowed to grow, the surplus words will cause the mission statement to be hard to comprehend. The best mission statement the author ever saw was one for an Accounting department. The mission was "to keep score."

Just to make certain that the statement is short, seek out and eliminate any "advertising"adjectives. Do the same for advertising adverbs. The mission statement tells what is to be done, not how to do it. Make sure there are no quality descriptors to tell the degree to which the mission is to be accomplished. The mission statement should be no more than one page long, even with business parameters. These two topics should contain no quality descriptors to tell how things are to be done.

SUMMARY

The mission statement is a more immediate starting point for the FA process than is the business description. Its purpose is to describe the heart of the business. The verb is very important because it describes the thrust of the business. There is a big difference in emphasis between "making a product"and "selling a product."

The mission statement tells *what* the business does. From now on, the logic process of FA is aimed at determining *how* to accomplish the mission. Step-by-step the team will decide the major goals, strategies, and operating rules by which the mission is to be accomplished.

5
Measures of Business Performance

INTRODUCTION

A mission can be very high sounding and idealistic. But is it realistic? This step of the Functional Analysis (FA) sets up a series of measures to ensure that the mission of your business is realistic. If you apply practical measures to the performance of a mission, then the mission becomes very tangible. You are forced to look at whether it is possible to determine the degree to which the mission will be accomplished.

The application of metrics is a sure way of assessing the true value of an idea. The practicality of the mission statement can only be judged to the degree that the performance of the mission can be measured. This is the first step of FA that might cause a reiteration of a previous step. The choice of practical measures of mission performance might cause you to change the description of the business or the business mission.

DICTIONARY DEFINITION

The definition given by *Webster* for the word *measure* is "any criterion for measuring or judging." And according to *Webster,* the word *performance* means "the degree to which any thing functions as intended." These two definitions will serve as a basis for this step of the study.

WORKING DEFINITION

The two definitions can now be put together to give a working definition of *measures of business performance*. The measures are defined as "criteria for judging the degree to which the business unit accomplishes its mission." This definition should be a simple idea for you to use to test your mission statement.

CRITERIA FOR MEASURING

The typical criteria for measuring the degree to which a business performs its mission are shown in Figure 5–1. They are of two types, one from the owner's point of view and one from the viewpoint of the product. The customer places emphasis on how well the product pleases him or her. The business is concerned about how well the product serves the customer and how effectively the product is produced and distributed. These concerns are labeled here as "product viewpoint."

The usual criteria are based on ownership interests. Some examples are:

- Profit as a percentage of sales
- The percentage of return on assets
- Sales volume in units or in dollars
- The percentage of sales as compared with that of total competition

These measures tend to be quite similar from business to business. They relate to the owner's viewpoint. They relate to profit more than they relate to the inherent nature of the business itself. Because of that, they are not really adequate to the purposes of FA. The measures from the viewpoint of the product are much more valuable as an aid to the analysis of the systems of the business.

There are three criteria that are based on the product itself:

- The quality of the product—the degree to which it serves the needs of the customer
- The timeliness with which the product is available to the customer
- The cost per unit of product

Unlike the measures from the viewpoint of the owner, these measures are likely to vary from business to business. They depend on the specific nature of the business. This is because they relate directly to the product, which is the heart of the business.

1. Owner viewpoint
   - Profit
   - Return on assets
   - Sales volume
   - Percentage of market
2. Product viewpoint
   - Quality
   - Timeliness
   - Cost per unit

Figure 5–1. Business measurement criteria.

EXAMPLES

An example of measures of mission performance for Hydroclean is shown in Figure 5–2, and an example for Gardentool is shown in Figure 5–3. There can be a number of different measures from the product point of view for any given product and business. For example, quality can be measured in terms of:

- Reliability, as in mean time between failures for a group of electronic components
- Life expectancy, as in the expected hours of life of a light bulb
- Utility, or suitability to the user's needs, as in the way information is represented on an engineering drawing
- The cost of warranty as a percentage of sales for an automobile
- The percentage of sold products needing repair within the first year of use, as in television sets
- The percentage of products that require replacement within the warranty period, as in garden hose

HYDROCLEAN MEASURES OF PERFORMANCE

Owner viewpoint

Profit as a percentage of sales

Percent return on equity

Dollar value of sales

Percentage of industry sales in the United States

Product viewpoint

Warranty as a percentage of sales dollars

Percentage of products needing repair within one year of sale date

Percentage of products shipped in time to meet customer requested date

Enterprise cost per unit of product sold

Figure 5–2. Hydroclean measures of performance.

- The ratio between the number of customer complaints and the volume of sales, as in missed deliveries of a newspaper.

There can also be different measures of timeliness for a given product and business. For example, timeliness can be measured in terms of:

- The percentage of products shipped on time, as in shipments from a mail-order nursery
- The percentage of products received on time, as in engines for installation in boat hulls
- The percentage of products ready to ship on receipt of customer order, as in warehouse shipments of washers and driers

GARDENTOOL MEASURES OF PERFORMANCE

Owner viewpoint

Profit as a percentage of sales

Percent return on equity

Dollar value of sales

Percentage of industry sales in the United States

Product viewpoint

Percentage of customer deliveries resulting in complaints

Percentage of line items out of stock at time of customer order

Percentage of orders shipped within three days of receipt

Operating cost per customer shipment

Figure 5–3. Gardentool measures of performance.

- The value of lost sales for lack of product availability, as in Christmas cards.

It is possible to measure cost per unit in terms of cost to produce from the supplier's viewpoint or in terms of cost to buy from the customer's viewpoint. Or it can be measured in terms of cost to own. Examples might include:

- The cost of selling activity per unit of product, as in door-to-door sales of encyclopedias
- The cost of production or procurement per unit, as in washing machines

- The total cost per unit of product sold, as in automobiles
- The cost to operate a bulldozer per hour
- The cost per cubic yard to dredge a channel with a dragline.

METHOD OF REPRESENTATION

The best method of representing the measures is in the form of a simple list like those in the illustrations. The measures should be separated into the two categories, owner and product, in order to lend emphasis to the product measures that will be more useful in later stages of FA.

All possible types of measure need not be represented. The intent is not to be exhaustive. It is better to have a few sound, practical measures than to have a long, confusing list of items that is not likely to be used.

ANALYTICAL METHOD

The simple steps of the analytical method are shown in Figure 5–4. Measures from the owner's viewpoint are found in annual reports, yours or those of other companies. In addition to the usual financial measures, you might think about the longevity of the enterprise and the performance of the business versus the competition.

The product measures come from two frames of reference. You want to look at the product from the customer's viewpoint. These measures are in terms of quality, reliability, availability, delivery time, price, and operating cost. You also want to look at the product from the point of view of the business itself. These measures are in

1. Owner measures in annual reports

2. Product measures from:

   customer's viewpoint

   business's viewpoint

Figure 5–4. Business measures analytical method.

terms of reputation, lost sales, warranty cost, replacement cost, stock availability, production, and sales cost.

Do not try to put dimensions on the measurements at this stage. This will be done at the next stage of the analysis. But make sure that the metrics you describe are truly measurable. If you cannot really put dimensions on the measures or cannot afford to make the measures, then they are of no use to you.

COMMON ERRORS

As in any stage of the FA process, there are some errors that might be made. A list of some of them should help you to avoid the pitfalls.

It is a mistake to place target numbers on the measures at this stage. Setting goals is the task of the next step in the analysis. If you try to go too fast, you are in jeopardy of bogging down in detail.

It is a mistake to define measures that are impractical to collect or to count. Subjective measures such as "customer satisfaction" are of no use. The measure must not be something that cannot be obtained without undue cost. Life of the product in the hands of the customer might be a very practical measure for draglines but not for can openers.

A common error is to place emphasis on change from the previous year, rather than to use an absolute measure of performance. This is often phrased as "percentage of growth over last year's sales." This sort of measure shows a lack of imagination and is inappropriate. It might be a specific objective for a specific year. But it is not a valid measure of ongoing performance of the mission of the business.

Do not waste time trying to obtain team agreement on a single measure for cost, quality, or timeliness. It is quite acceptable to have several alternatives for each type of measure, so long as they are practical to apply. On the other hand, you can define too many measures. The very volume can cause confusion. Too much detail can make it difficult for the team to concentrate on that which is truly important.

Do not forget the customer's viewpoint. It is more important than the viewpoint of the business when it comes to assessing the performance of the mission.

CAUSES AND CURES

A common cause of error comes from confusing the definitions of *mission, strategy, goal,* and *objective*. Please accept the definitions in this book to avoid that confusion. These terms are defined in each chapter where they apply. Above all, do not let the team fall into a debate about the meaning of the terms, which would waste many hours and could delay the project by days.

Another source of error is in forgetting the viewpoint of the customer or in presuming that you already know it. It never hurts to talk to the customer or, better yet, to listen to him or her. This will be even more true at later stages in the analysis when you measure the performance of the missions of departments that provide products for internal use.

It is a mistake at any time to give in to shallow thinking. This is another area in which the ability of the team leader is stretched. The leader must test the depth of thinking of the team by asking them why they recommend each measure. If a measure gives no practical value, it should be rejected. FA is not just an exercise to "get through." The team must also evaluate the practical aspects of how each measurement is to be obtained. For example, it is more cost-effective per unit to maintain sales records of tractors than it is to do so for washing machines.

The last cause of error is a natural tendency to make time-frame comparisons: "Better than last year." Remind yourself that you are defining measures of the *ongoing* operation of the business from day to day.

QUALITY CHECKLIST

A quality checklist is shown in Figure 5–5 to help you to critique the finished document from this stage of the study. There must be at least one measure from the owner's viewpoint. You will probably want more measures than this, especially if the business is publicly owned. There must be at least one measure from each type of product viewpoint. There must be at least one for quality, one for timeliness, and one for cost per unit..

There should be no more than five measures from the owner's viewpoint. You can get carried away with a good thing. There

1. At least one from the owner viewpoint
2. At least one from each product viewpoint
   - quality
   - timeliness
   - cost per unit
3. No more than five from owner viewpoint
4. No more than two from each product viewpoint
5. Product measures from customer perspective
6. No dimensions on the measures
7. Realistic, affordable measures

Figure 5–5. Business measurement quality checklist.

should be no more than two measures from any single product type of viewpoint, and there should be at least one of each of the three types of product measures from the customer's frame of reference. You probably will also want at least one from the company's frame of reference for cost per unit.

There should be no dimensions on the measures, since these come with the goals in the next chapter. And the measurements should be practical ones. You should be able to say how each can be accomplished without undue expense.

SUMMARY

The measures of mission performance are the keys to the day-to-day operation of the units of organization within the business. Each department within the business will be driven, directly or indirectly, by the measures of performance of the business as a whole.

The mission and the measures are enduring. They will not change much from year to year. That is why they are different from objectives; they describe the very nature of the business.

The dimensions on the measurements come from the business goals that are defined in the next chapter. The annual planning of short- and-long term objectives is not a part of the FA process.

In the next chapter, you will assign numeric values to the measures. You will determine just how well the business must perform in order to accomplish its mission. The next question after that is the hard one: How should the internal operation of the business be conducted in order to satisfy the measures? That is the very essence of FA.

6
Business Goals

INTRODUCTION

Goals are not as unchanging over time as a mission is, but they are the things that drive business strategies. At least, that is the flow of logic in functional analysis (FA). Business strategies, on the other hand, are statements of organizational behavior aimed at achieving specific goals.

Nor are goals as changeable as business objectives, which are expressions of intent to improve the performance of the business over a time period of from one to five years. Goals should be long term. They should direct the activities of the business for time spans of at least five years.

The setting of goals is the next logical step in analyzing how the business is being operated and what changes might be appropriate. For example, it makes a big difference whether a goal is to fill customer orders within three days or to fill them within three weeks. The difference is in who holds the inventories, the business itself or those who supply it.

DICTIONARY DEFINITION

Webster defines the word *goal* as "an end toward which one directs his exertions."

WORKING DEFINITION

For the uses of FA, *goals* are "the specific dimensions that the management of the company sets on the measurements of day-to-day performance." The goals are the numbers assigned to each unit of measure. The goals provide specificity to the measures so that they can become the basis on which realistic strategies can be developed.

1. Target number

2. (Target date)

Figure 6–1. Dimensions for goals.

DIMENSIONS

Refer to Figure 6–1 for the types of dimensions that can be assigned to measures in order to change them into goals. The usual choice of dimension is a target number for each measurement, as in a specific percentage of profit or a specific number of days within which to respond to a customer's order.

Sometimes a date might be assigned for achievement of the target dimension if there is an intermediate goal, as there might be for the early years of a new business. For example, the goal might be 0.1 percent manufacturing scrap with a three-year intermediate "goal" of 0.5 percent. These intermediate goals are indented in order to place emphasis on the end goals.

EXAMPLES

An example of goals for Hydroclean is shown in Figure 6–2 with dimensions assigned to the measurements that were defined in the last chapter. Figure 6–3 shows an example of goals for Gardentool. Note that there are intermediate goals in both examples.

METHOD OF REPRESENTATION

The goals are the same measurement statements from Chapter 5, with the simple addition of specific dimensions to the measures. A goal might be stated as several progressive steps if there is need for one or more intermediate objectives, to build the business up to the desired level of performance over a period of time.

ANALYTICAL METHOD

Figure 6–4 lists the steps involved in the development of goals. Top management must set the targets, goals, for the business. These

HYDROCLEAN BUSINESS GOALS

Owner viewpoint

A profit of 8 percent of sales

12 percent after tax return on equity

$300,000,000 annual sales in 1980 dollars

35% of total U.S. industry sales

Product viewpoint

Warranty less than 1 percent of annual sales

Less than 3 percent of products needing any sort of repair within one year of the sale date

95 percent of products shipped in time to meet customer requested date

$100 enterprise cost per unit sold in 1980 dollars

$120 enterprise cost per unit sold in 1980 dollars — by 1995

Figure 6–2. Hydroclean business goals.

goals should, however, have some input from the FA team. After all, it is the team that must determine how the goals are to be accomplished. Most often, the team recommends goals to the sponsor of the FA project.

What does competition do? One thing to consider when setting your own goals is to evaluate what goals the competitors are likely to have set or what they have actually advertised as goals. The goals must have numbers, if possible, as in the examples. But sometimes goals can be set with implied dimensions. An example might be: "to be the leader in industry sales within the United States." This sort of goal is generally less clear and should be avoided if possible.

GARDENTOOL BUSINESS GOALS

Owner viewpoint

5 percent profit as a percentage of sales

13 percent return on equity

$200,000,000 annual sales in 1980 dollars

20 percent of industry sales in the United States

Product viewpoint

Less than 1 percent customer complaints as a percentage of orders

Less than 3 percent of line items out of stock at the time of customer order

Less than 6 percent of line items out of stock at the time of customer order by 1993

90 percent of orders shipped within three days of receipt of customer order

Operating cost less than 2 percent of the total value of customer orders

Figure 6–3. Gardentool business goals.

The goals must be achievable in five to ten years. Otherwise they are unrealistic. Since the goals will drive the business strategies, it is foolish to set unrealistic goals that in turn will lead to impractical strategies and costly operating practices. If the ongoing goals are not achievable in a reasonable time frame, then intermediate goals must also be set to guide the development of operating strategies.

Always test the goals for realism. Do you know how to attain the target that is set with each goal?

1. Goals set by top management
2. Considering competition
3. With dimensions
4. Achievable in five to ten years
5. Realistic, achievable

Figure 6–4. Business goals analytical method.

COMMON ERRORS

The most common error is to set unreasonably high goals. It is good to stretch yourself; but it is foolish to intentionally stretch beyond your reach. Another error is to set goals that are incompatible. This can happen when the owner viewpoint and the product viewpoint are not compared. It does no good to meet one specific product goal if that causes you to fail to meet other product goals or the owner's goals.

CAUSES AND CURES

One cause of unachievable goals is a lack of appreciation of the effort and cost it might take to attain them. If you make that mistake at this stage of FA, it should become clear during the strategy setting stage. It will then result in the need to reiterate through the process of setting goals.

Sometimes it is hard to obtain top-management input into the setting of goals. The goals and the resulting strategies need to be discussed and evaluated from both the ownership viewpoint and the product viewpoint. If top management does not have the time to set goals, then the team must offer goals to them for their approval.

Another cause of unreasonable goals is an inability to make the necessary operational changes in a reasonable time frame. In this case, the answer is to set intermediate goals as well as final goals.

1. Dimension on each measurement
2. Long-term goal for each short-term goal

Figure 6–5. Business goals quality checklist.

QUALITY CHECKLIST

A quality checklist for evaluating your goals is shown in Figure 6–5. The evaluation is very simple. There must be a dimension, a target number, on each measurement. Otherwise, it is not a goal. There must be a long-term goal to accompany any short-term goal that is qualified with a date. If you do not do this, you have set an objective, not a goal. Remember to indent the short-term goals in order to lend emphasis to the long-term goals.

SUMMARY

This step of defining business goals is an essential one before you try to set business strategies. It is a very short step but an important one none the less.

The business must have valid goals for performance, or else the business strategies will not have any practical basis. They will most likely be idealistic and impractical. For example, today there is a general move of U.S. industry to the concepts of JIT and CIM. You need to determine if these new ideas lead to the accomplishment of the business goals of your company. It is important to decide if the business goals of your company should be revised because of what these concepts make possible. The last thing you want to do is to adopt JIT or CIM just because they are popular.

The use of computers should serve a business purpose. The use of computers is not in itself a valid business purpose.

No matter how realistic you think your goals are, the development of strategies will almost certainly cause you to rethink the practicality of some of the goals. That is fine. You have the ability to go back and to change any goal. If you do so, you will then have a clear understanding of why it was necessary.

7
Business Strategies

INTRODUCTION

The definition of the word *strategy* is much debated since it seems to have many different meanings to people. It is often used interchangeably with the word *goal* and at times is used interchangeably with the word *objective*. There are also people who find it difficult to differentiate between *strategies* and *tactics*.

If you permit your team to debate the meaning of the word, you will set back progress on the functional analysis (FA) study for several weeks. For the purposes of the FA, it is recommended that you accept the definition given in this chapter. Whether or not you agree with it, it will serve your analytical needs.

Two strong forces that drive the day-to-day operation of a business are the measures of performance and the strategies. For this reason, it is most important that the FA team spend serious time and attention on this phase of the study.

DICTIONARY DEFINITION

Our good friend *Webster* says that a *strategy* is a "method, plan, or stratagem to achieve some goal." This definition is not as good as it might be. It is not supposed to be proper to define a word by using a form of the same word.

WORKING DEFINITION

There is a definition of the word *strategy* that has proved to be useful for FA. It is "a statement of a principle of operational behavior or judgment that serves as a guide for day-to-day decisions." It might be added that these guides are needed so that the goals for accomplishment of mission will be met. A strategy is a way of reaching a goal.

Strategies must be supportable with a rationale. A strategy is good, not because it sounds good but because it is better than other sensible alternatives. If a strategy sounds like "motherhood," that is probably what it is. If there is no reasonable alternative, then the strategy statement is most likely to be trite—and not useful or needed.

One good test of the sense of a strategy is to define one or more opposing strategies. Then seek to find a reason why each opposing strategy is more or less preferable. If you cannot phrase an opposing strategy that makes sense, the one you started with is probably not needed.

STRATEGIC TOPICS

Since strategies are to serve as guides to help meet goals, they should deal with the means by which those goals are to be reached. Some possible topics for business strategies are shown in Figure 7–1. They deal with a number of business needs:

- *Pricing*. In order to obtain a desired level of market share and to attain a profit goal
- *Differentiation*. What will make the product more attractive to the customer than the products of the competition
- *Segmentation*. To corner the market among a specific set of customers
- *Distribution*. To manage the delivery of the product to the end user
- *Financing*. The method of providing the capital needed to support the business
- *Procurement*. To manage the acquisition of materials for production
- *Manufacturing*. To manage the actual production of the product you sell
- *Finished inventory*. How much inventory of final product to have on hand and how to balance its costs with responsiveness to customer orders
- *Product support*. What will be done to ensure the quality image of the product when it is in the hands of the customer
- *Product change*. To keep a degree of production stability while responding to market demand for new products

1. Pricing
2. Differentiation
3. Segmentation
4. Distribution
5. Financing
6. Procurement
7. Manufacturing
8. Finished Inventory
9. Product Support
10. Product Change
11. Technology
12. Quality

Figure 7–1. Strategic topics.

- *Technology*. To balance risk versus progress in the introduction of change to the product or to the method of production
- *Quality*. To put in place a method to ensure that the final product meets target levels of reliability and performance.

This list of examples is a bit long but is in no way exhaustive. You can, and will, find many other examples in your own business.

Note that strategies as often as not are a means of striking a balance among potentially conflicting goals. Cost goals can conflict with service goals; risk goals can compete with progress goals. The strategy statements that your team selects will surely cause conflict among the departments of the company. The resolution of those conflicts will occupy a large amount of your time in future stages of your FA study.

EXAMPLES

The format for strategies is shown in the example for Hydroclean in Figure 7–2. An example for Gardentool is shown in Figure 7–3. Take a moment to think about the impact of some alternative

HYDROCLEAN STRATEGIES

1. Prices will be within 5 percent of leading competitors.
2. Products will be aimed at home use.
3. The market place is the United States.
4. Distribution and product support will be through department store chains.
5. Product will be given the label of the department store chain when requested.
6. Frames, drums, panels, and wiring harnesses will be produced in-house.
7. Suppliers will be sole source for the items they supply.
8. Material procurement and production will be based on Just-in-Time principles.
9. Product change will be annual.
10. No new technology will be introduced until it has had two years of successful industrial use.
11. User instructions will be minimal and a part of the product itself.
12. Quality is targeted at ten years of life without repair from the time the unit is installed.

Figure 7–2. Hydroclean strategies.

GARDENTOOL STRATEGIES

1. Prices will be no higher than competitors' for major items.
2. The product catalog will provide one-stop shopping for yard and garden retailers.
3. The marketplace is the Southeast.
4. There will be no retail sales.
5. The only products with Gardentool label are seeds and plants.
6. Inventories are to be maintained on the basis of product perishability or obsolescence.
7. Product support will be limited to replacement parts.

Figure 7–3. Gardentool strategies.

strategies for Hydroclean in the area of management of production inventories. These examples show how important the choice of strategies can be.

A first possible strategy could relate to the manner in which the company might choose to obtain component parts for the product:

1a. Purchase all the production materials; make no parts
1b. Buy parts but manufacture all fabricated assemblies in-house rather than buying them

A second possible strategy might relate to how the final product is to be put together:

2a. Assemble all final product in daily batches to keep a level of finished inventory on hand
2b. Assemble each unit of finished product only as a customer order calls for it

There are four possible ways in which the business will operate, depending on which strategy it chooses from each of the alternate 1 and 2 statements. The pair that is chosen can have four very different impacts on how the procurement and production processes are to be operated:

The choice of 1a and 2a would result in daily receipt of regular quantities of finished assemblies in accord with a master production schedule. These would most likely be produced by the supplier each day in a batch and then delivered each night to the product assembly area.

The choice of 1a and 2b would create a need for JIT manufacturing capability at the supplier. Or it would make it necessary to keep a buffer of fabricated assemblies for the assembly process. The buffer would probably have to be held by the supplier and be delivered each day.

The choice of 1b and 2a would make it necessary to store only parts and raw materials in the plant for assembly use. It would leave the degree of JIT manufacture of assemblies optional. Ask yourself what value JIT manufacture would bring.

The choice of 1b and 2b would require the company itself to keep an inventory of parts and raw materials for assembly. It would also require an inventory of finished assemblies or the introduction of JIT manufacturing to supply the assembly area. This would probably result in the need for computer control of material movement and manufacture.

The first pair of strategies would allow the production operation to be fairly simple and would minimize the need for use of a computer. The last pair of strategies would complicate the production process a great deal. This last pair would probably make it necessary to use a computer in order to respond to changes in the volume and mix of work from day to day.

METHOD OF REPRESENTATION

The strategies are listed as simple, complete statements on a single page. For the sake of ready reference and comprehension, they should be limited to no more than a single page. From 6 to 15 statements is a good number of strategies. If you try to list too many, there is a tendency to list strategies that really belong at the department level. These departmental strategies are the topic of a later stage in FA, in Chapter 11. A business should be able to operate with about ten key strategies, since it is difficult to keep more in memory.

Each strategy statement should be supplemented by a rationale to tell why it is better than the alternatives. Again, the statement of rationale for each strategy should be no longer than a single page. Examples are shown in Figures 7–4 and 7–5. The rationales are not needed as reference documents for frequent use. They are to be used as aids to help others understand why each strategy is important and why it is preferable to alternatives. Strategies are frequently challenged. The statement of rationale helps to answer these challenges in an organized, consistent manner.

ANALYTICAL METHOD

A series of recommended steps in the analytical process for business strategies is shown in Figure 7–6. Top-management participation is important in the development of strategies. If top management is not available to help develop the strategies, they should at least review the strategies that are recommended by the team.

The development of the strategies begins with a detailed review of what business you are in and why. The work you have done up to this stage has prepared the materials you need for this task. Consider the strong points of your company. What current strategies seem to be in place? They are often undocumented, but they are easy to find if you ask how the company seems to prefer to do business. Consider any weak points the competition might have. This might provide clues as to where you should emphasize a current strategy or develop a new one. Consider at least one alternative to each proposed strategy. The first ideas are not always the best. Besides, if you do not look at alternatives, you will not have much of a rationale for why you have selected a specific strategy. "We have always done it that way" is not a valid reason for anything except going out of business.

Consideration of alternate strategies might lead to other topics relevant to the business. For example, the choice of 1b and 2a in the Hydroclean example leaves open the question of whether or not the production of manufactured assemblies should be done JIT. A deci-

HYDROCLEAN STRATEGY RATIONALE

1. Prices will be within 5 percent of leading competitors.

The intent of this strategy is to emphasize the strength of our company and the reputation of our product for quality, serviceability, and reliability. If the primary customer motivation for purchasing a Hydroclean washer or drier is price, we are put into a competition for cutting corners on those product characteristics that have differentiated us from the competition.

We do not intend, on the other hand, to pursue those customers who buy on the basis of high price or snob appeal. We will continue to emphasize the fact that our product can be installed and operated reliably for at least ten years from the time of purchase. Its continued reliable performance should be taken for granted by the customer, like a furnace or a telephone.

Because of the growth in apartment living and of utility rooms on the main floor of houses, the need for reliability and the avoidance of water overflow or leakage has become paramount.

By the same rationale, we must restrict our product outlets to chains with a high-quality image.

Figure 7–4. Hydroclean strategy rationale.

sion is needed on this topic, either at this stage or at the time when you develop strategies for each department.

Here are some alternatives to look for when defining your own set of strategies:

- *Seasonality*. Buy rather than make, or make seasonal alternatives such as mowers and snowblowers, or sell worldwide to take advantage of global seasons

GARDENTOOL STRATEGY RATIONALE

1. Prices will be no higher than competitors' for major items.

Our intent is to become the sole source supplier for our customers - retailers of yard and garden products. There are two major factors that influence our customers, price and service. We must distinguish ourselves on the basis of service and the convenience of doing business with us. But we must expect that the customer will still compare prices with our competition. It is easiest for the customer to compare prices on major items such as garden tractors and rototillers. Since both we and the competition purchase these items from the factory, the difference in the price we quote is a direct measure to the customer and to us of how effectively we operate.

Figure 7–5. Gardentool strategy rationale.

- *Promotion.* Sell product or sell systems using the product; or sell directly to the end customer or indirectly to retailers who in turn sell to the end customer
- *Leverage.* Be the low-price supplier or differentiate the product in order to command a higher price.

Seek a consensus on each strategy and its rationale, first from the team and then from top management. It is important that the different departments buy in to the strategies, since they will be affected by them.

COMMON ERRORS

You can develop too many strategies or too few. If you develop too many, the team will lose perspective on which are the most important ones that are needed to direct the business. If you have too few,

1. Top-management participation
2. What business are we in?
3. Why are we in this business?
4. Company's strengths
5. Competitors' weaknesses
6. Alternatives
7. Rationale - why?
8. Why not the alternative?
9. Multidepartmental consensus

Figure 7–6. Business strategies analytical method.

they do not provide enough direction to the team for the remainder of the project.

A second common error is to choose strategies that are without real meaning. They are trite and meaningless as in "keep production costs to a minimum."

CAUSES AND CURES

One cause of poor strategies is an attitude that the development of a set of strategies is just an exercise. "It is something that must be finished as soon as possible so that everyone can return to business as usual." This error will not happen if there is an active sponsor and if the team is well constituted. It can also be avoided if time is set aside for team meetings, away from the interruptions of daily work.

Another cause of poor strategies is shallow thinking or statements with no logic to back them up. The team leader can avoid this by forcing a debate of rationale, involving a number of people of different responsibilities. This problem is also minimized if top management is involved in the review and approval of the final strategies. If there are a lot of "but's," the strategy is probably ill conceived. If a strategic statement is only valid under a list of special conditions, then it is not a true strategy. Look harder to find a statement that can have a broad impact.

One final cause of poor strategies is the person who cannot think in broad terms. This is the person who sees all things only from his or her own departmental point of view. The team must include imaginative people who are able to consider the needs of all areas.

QUALITY CHECKLIST

Figure 7–7 shows a list of factors to use in order to validate the strategies before the team has finished with them. Make sure that the statements are simple and to the point. If they are long and rambling, they will just confuse people and will not be directive.

Review the strategies to make sure they reflect the viewpoint of the business and not just that of individual departments. If too many of the statements seem to be specific to individual departments, they do not belong on this list. Set them aside for now and refer to them again when you reach the step described in Chapter 11.

Keep the number of strategies between 6 and 15 so that they can be remembered, and followed, in later stages of the study and also in

1. Simple statements
2. Not departmental
3. No less than 6 strategies
4. No more than 15 strategies
5. Fits on one page
6. A rationale for each
7. Consensus

Figure 7–7. Business strategies quality checklist.

the daily operation of the business. Keep the total list of strategies to a single page. If it becomes longer, the impact of the statements will be diluted. You will begin to write a book that will never be read or implemented.

Make sure that there is a page of supporting rationale for each strategy. If you do not have it, you can be sure that in the future someone will misinterpret a strategy. Or someone will challenge a strategy without anyone from the team being available to explain why it is important.

Make sure that there is a consensus and a buy-in from each member of the team for each strategy. And make sure that there is agreement from top management or at least from the sponsor.

SUMMARY

The business strategies are the guides to the next steps of FA in which the team will begin to develop the departmental missions and strategies. This is very important, since strategies can have a profound impact on the individual departments. In the example cited earlier, there might be a decision not to manufacture at all, thus obviating the need for a department. This is what Firestone decided to do.

At this point in FA, there is an absolute need for consensus on the content of all the documents that have been developed so far. Review the previous steps, if necessary, in order to obtain it. And seek agreement from the sponsor and from other members of top management, if possible.

The thought process is now ready to go from the broad viewpoint of the business as a whole to the departmental level, with the assurance that there is a common thread of logic to guide the individual departments.

Even though the importance of the strategies has been emphasized, it is well to remember that they are guides and not absolute laws. In daily operations, there will be an occasional need to act contrary to a given strategy. But this should be the exception and not the rule. If the exception occurs too often, the strategy was ill conceived.

8
Department Mission

INTRODUCTION

The statement of the mission of a department within a business is a representation of what that unit of the business does. That does not seem at first glance to be of great value to functional analysis (FA), since the people in the department already know what they do. But it is of very great value. It is an analytical step, since it is a way to begin thinking about the business in a new and structured way.

The mission statement has the potential of being of unusual significance, because the previous FA steps could well have caused a need to change the mission of a given department. A portion of a business might have actually had little freedom to define its mission in the past but might now see a new purpose. At any rate, the development of a mission statement can be quite revealing. Often the mission of a department has been tacitly understood but never documented; and most likely it has never been evaluated to see if it really describes what the department does—or should do.

DICTIONARY DEFINITION

As usual, it is best to start with definitions. *Webster* says that a *department* is "a division of a complex whole or organizational system." The word *mission* was defined in Chapter 4 as "the sending of an individual or group by an authority to perform a specific service." The broad use of the term *mission* by industry has led to the evolution of a new meaning. A mission is generally accepted to be the function or ongoing activity that a unit of business is responsible to perform. The mission is a statement of what the unit must do.

WORKING DEFINITION

In FA a *mission* is the responsibility to deliver a product to a customer—to the person who needs to use it. The product can be any one of three types—material, information, or service.

If the product of a department cannot be identified, it suggests strongly that there is no clear mission and that the unit of business has no reason to exist. The same is true about the need to identify a customer. If there is no customer, then there is no evident reason for the department to exist. This is really something that might motivate the head of a department to define his or her mission!

If the product of a department does not go to the end customer of the business, one must ask if it in any way adds to the value of the end product. If the product of the department never passes out of the business itself, then it must add value within the business. That value could be in the form of enhancing the end product, of reducing the cost of producing it, or of helping in some way to move it into the hands of the customer. But if the product of the department does not add value, it should not be produced. And a question must be asked as to the need for the existence of the producing unit.

COMPONENTS OF A MISSION STATEMENT

The component parts of the mission statement of a department are the same as for the mission statement of a business. To save you the need to refer to a previous chapter, they are shown in Figure 8–1.

1. Do something
2. With a product
3. (From a source)
4. (Through an agent)
5. For a customer
6. (For a purpose)

Figure 8–1. Components of a mission statement.

Do something. The department is responsible to do something. It might produce, report, schedule, plan, and so forth.

With a product. It must do its thing with a product of some sort. The product might be material, as in piece parts. It might be information, as in a market report. It might be a service, as in the hiring of new employees.

(From a source). The product or its components might come from some special source that is important enough to be mentioned. For example, a market report might be developed from sales estimates from the field force, or it might be extracted from trade publications. This portion of the mission statement is shown in parentheses because it is optional.

(Through an agent). This part of the mission statement is also optional. The product might be contracted for by the department but actually produced by an outside group. An example would be the installation of new machinery for the shop floor. Production Engineering could plan it, buy it, and contract for it to be installed by a third party.

For a customer. It is not always obvious who the customer is. Or there might be multiple customers. In these cases, it is best to be clear and to mention the user or users of the product.

(For a purpose). This is another optional part of the statement. The use to which a product is to be put has a lot to say about the character of the product. If spare parts are to be used by a Service Department, they might need to be packaged and stored in a different manner than if they are to be sold to the end customer.

EXAMPLES

Figure 8–2 shows what a typical statement might be for the mission of a Manufacturing Department. Like the mission statement of a business, it is short and to the point. Figure 8–3 shows a mission statement for Materials Handling. And other examples are for Purchasing, in Figure 8–4; for Production Scheduling, in Figure 8–5; for Order Entry, in Figure 8–6; and for Assembly, in Figure 8–7.

The reason for the number of examples is to build a basis for later examples of department strategies and conflicts. The departments in these examples have a great deal of mutual dependency in terms of internal products.

MANUFACTURING

Mission: To produce major components of Hydroclean products

Figure 8–2. Manufacturing mission.

MATERIALS HANDLING

Mission: To move materials to the point of need

Figure 8–3. Materials Handling mission.

PURCHASING

Mission: To buy the materials needed for the production process

Figure 8–4. Purchasing mission.

PRODUCTION SCHEDULING

Mission: To set the rate of production of final product

Figure 8–5. Production Scheduling mission.

ORDER ENTRY

Mission: To balance customer demand with final product availability

Figure 8–6. Order Entry mission.

ASSEMBLY

Mission: To assemble final product from components and subassemblies

Figure 8–7. Assembly mission.

No more reference will be made to Gardentool, which is a very simple business compared with Hydroclean. From this point on, all examples will be from Hydroclean.

METHOD OF REPRESENTATION

A simple statement of limited length is the best way to record the mission of a department. It is not necessary to use a lot of descriptive words, which will just waste time and space. Brevity and clarity are as important here as they were for the business mission—and for the same reason.

The mission statement is only the first step in defining a department. It will be expanded into a full document in later steps of FA.

Business parameters are not needed at this stage, as they were for the business mission. This need will be handled in a different way in later steps.

ANALYTICAL METHOD

The steps in the analytical method are shown in Figure 8–8. The first step is to identify the departments—not a trivial task. The term *department* means different things in different companies. The examples in this chapter show the types of functions that are usually called departments. But the division of the business into smaller units need not be done along strict lines of organization. Consider a department to be a unit of the business that produces a significant product. If in doubt about a unit, call it a department.

If only a subset of the business is to be studied, then only the relevant departments should be included.

Identify the product of the department. If the unit is the source of a number of products, identify the most significant product. It does not matter if the product is material, information, or service. You

1. Identify the departments
2. Identify the product
3. Identify current practice
4. Use action verb
5. Incorporate business strategy
6. Verify with department head

Figure 8–8. Department mission analytical method.

need to find out what each department does. This might or might not already be documented. Most likely, it is not. You do not need to go to any great detail at this step. But you need to be specific about the manner in which the department performs its mission. Does it contract with another business for the product; or does it make it itself? For whom is the product intended? You need to be able to supply the information listed in Figure 8–1.

The verb is critical, just as it is in the business mission. Does the department produce or manufacture, sell or wholesale, design or advise; does it distribute? The verb tells *how* the department performs its mission.

If yours is a new business, or if it has been changed from the past by a change in business strategies, then you need to document only the new mission. Always review the final result with the department head to ensure that he or she agrees.

Other possible changes in the mission will be developed as a part of the strategies step in Chapter 11.

COMMON ERRORS

This step of FA is subject to the same errors as is the business mission step. The most common error is to write too long a statement of mission. As before, keep it to the point—short and simple. The second type of error is the use of the word *and* too easily. It certainly adds to the length of the statement. You do not have to be comprehensive. There is plenty of time for that in later stages.

The final error that adds length to the statement is the use of adjectives and adverbs to show a level of performance or quality. They not only add to the length of the mission statement; they cause it to be hard to read and understand.

CAUSES AND CURES

One cause of long mission statements is the attempt to "word smith" for advertising effect. It is much better to try for clarity and brevity of image. Avoid all adjectives, adverbs, and conjunctions that are not vitally needed. This is another reason why so many examples are included in this chapter. It is hoped that the examples can show that it really is possible to be terse but accurate.

It is common for people to want to say how the mission is to be performed. It is not unusual to see statements like "to be the low-cost manufacturer of premium widgets." This is, at best, a goal or, at worst, propaganda. You need to say what the department does. You might describe the medium through which it is done, but do not ever say how well it will be done. Not here. That comes later.

It is not unusual for people to attempt to strategize or to try to make decisions too early in the process. Take notes of any ideas for strategies and save them for a later step in Chapter 11. And some people cannot resist the desire to be all-inclusive. Remember that you are to seek clarity at this point, not completeness.

Often there are different types of products that are provided by a single department. These can be products of the subfunctions of the department. An example of a subfunction and a secondary product might be the report made by the assembly line of the number of final units of product built in a day. These subfunctions and their products will be defined later, in Chapter 10. Restrict your effort in this step to major products only.

QUALITY CHECKLIST

Figure 8–9 shows a checklist for you to use to test the quality of the statement of mission for a department. A mission always begins with the word *to*. And that is followed by the verb. The mission must be stated in 15 words or less. If it is not, it will be harder to read and harder to recall to mind when needed.

1. Begins with the word *to*
2. Less than 15 words long
3. No "advertising" adjectives
4. No "advertising" adverbs
5. No quality descriptors

Figure 8–9. Department mission quality checklist.

The statement should have no "advertising" adjectives to tell how well it will be done. The purpose of the statement is to inform, not to impress. The same is true about "advertising" adverbs. There is never a need for them. And there should be no quality descriptors of any kind. The mission should be simple and to the point.

SUMMARY

The mission statement is the immediate starting point for the detailed steps of FA. From now on, the logic pro[illegible]s is aimed at how to make the departmental mission happen[illegible]jectives and adverbs will begin to be of importance as you define measures and strategies for the departments.

From these next steps, you will begin to find areas of conflict among the departments. These are points where the units tend to work at cross purposes because of differences in their strategies and measures of performance. The systems of the business will be clarified as you seek to find ways to reduce the points of conflict.

9

Measures of Department Performance

INTRODUCTION

The measures of performance of a department's mission are the things that drive the day-to-day work of the unit. They are the chief motivators of operational behavior. They are the key ideas that make the department run for the benefit or disservice of the business as a whole.

Another motivator of behavior is the personal career ambition of key personnel within the department. This will not be dealt with to any depth within this text. FA deals with functional activity, not with personalities. Anyone who has not seen selfish behavior at high levels of management has not had broad exposure to the pressures of the business world. Very few people act with the best interests of the whole business in mind. To expect that they do is to be naive. People tend to act in a manner that they feel is expected of them by their boss. They act provincially.

In my book *Management Information Systems* there is phrased a law of human behavior that can be stated as: "We do as we are measured." There are a number of examples of this law in this chapter and in later chapters of this book. It is a human analogy of the carrot and stick story about the donkey. We respond to rewards, and we shun those things that bring pain.

The definitions of measures, and goals, of department performance should be developed with this principle in mind. *Measures* are the means to ensure that people will actually perform their duties in the desired manner, for the benefit of the enterprise.

MEASURES ARE NOT PERMANENT

It is important to note that measures can and should change over time as conditions change. There can be changes in such things as the business climate, the rate of technical progress, the needs of the customer, the actions of competitors, and so on.

The measures that you define in this step will also be subject to question and to change. They will need to change as the environment changes. But you will also find need to change them over the next few weeks or months as you develop new ideas in later steps of functional analysis (FA).

WORKING DEFINITION

The *measures* are criteria to help judge the degree to which the unit is able to fulfill its mission. They gauge the degree to which the unit provides the product to the satisfaction of its customer's needs.

CRITERIA FOR MEASURING

The bases for measuring the performance of a department are the same three by which the business is measured from the point of view of the product. They are listed in Figure 9–1—quality, timeliness, and cost per unit.

These measures are the drivers of individual and unit performance, but they are also the key causes of conflict between the units. This should begin to become clear in the next few pages. You will see how the measures of performance in one unit can cause it to deliver product that does not satisfy what some other unit needs in order to meet its own measures of performance. If this sounds strange, think of a Purchasing Department driven by a tough measure of cost per unit and a Manufacturing Department driven by a

Product viewpoint

1. Quality
2. Timeliness
3. Cost per unit

Figure 9–1. Department measurement criteria.

low scrap target. Purchasing will strive for low prices, possibly at the expense of Manufacturing's need for quality.

The criteria for measuring department performance relate to the product of the department:

- The quality of the product as measured in terms of its value to the user
- The timeliness of delivery of the product to the user relative to when he or she needs it
- The cost of the product, either in terms of cost per unit of production or in terms of the price paid by the user

There is no way to sum up these measures into a single measure. They must be applied as individual measures, each of which must be satisfied. Satisfying each measure is vital to proper performance of the mission of the department. If any one of the measures is not met, the product is a reject, and the mission is not accomplished. If quality is bad, the product is not usable, and the user must seek some other product to meet his or her needs. If the product is not available on time, the user will be late in performing the function for which the product is needed. The user will miss his or her own timeliness target. If the cost per unit is too high, and if the user pays for the product, then he will seek a less expensive source. If the cost per unit is too high, and if the supplier pays for the product, then the supplier will seek to reduce the cost, possibly to the detriment of quality and timeliness.

MEASUREMENT VIEWPOINT

The measures should reflect the customer's needs if the mission is to be accomplished as it should be. After all, the customer is the only one who finds real value in the product. Having said that, it needs to be added that the supplier also finds value in the product. It is the product that provides a job for the people in the supplying department.

But the third measure, cost per unit, need not be from the customer's point of view unless he or she actually pays for the product and has the option to seek other sources. If the customer does not pay for it directly, then the cost target is set by the chain of command over the supplying department. How much can it afford to spend to produce the product?

EXAMPLES

A number of examples of measures of department performance are shown in Figures 9–2 through 9–7. These examples relate to a group of departments that will be the source of reference throughout the rest of this book. These departments are:

- Manufacturing, as shown in Figure 9–2
- Materials Handling, as shown in Figure 9–3
- Purchasing, as shown in Figure 9–4
- Production Scheduling, as shown in Figure 9–5
- Order Entry, as shown in Figure 9–6
- Assembly, as shown in Figure 9–7

There are different rationales for the choice of measures for each of the departments. Each unit has its own mission, its own product, and its own customers. The measures can drive any one unit into actions that are not in the best interests of the other units. Thus, these departments will begin to serve as examples of the conflicts that are caused by the choice of measures of performance.

MANUFACTURING

Mission: To produce major components of Hydroclean products

Product: Major subassemblies

Customer: Assembly, Field Service

Measures: Less than 0.3 percent of production scrapped in production, at assembly, or in the first year of use

Less than 1 percent of assembly delay per week for lack of manufactured components

Cost per unit 10 percent less than delivered cost of purchased alternative

Figure 9–2. Manufacturing measures.

MATERIALS HANDLING

Mission: To move materials to the point of need

Product: Delivered production materials

Customer: Manufacturing, Assembly

Measures: Less than one error per thousand deliveries, including misdeliveries and damaged materials

Less than 1 percent delay in weekly production time in Manufacturing or Assembly due to late delivery of available material

Material handling cost less than 3 percent of total production cost of product

Figure 9–3. Materials Handling measures.

METHOD OF REPRESENTATION

The document will be expanded bit by bit in a logical pattern over the next few steps of FA. Three new topics are added as a result of the work of this step. The second item in the document is the main product or products of the department. Remember that the product can be material, information, or service. The third item in the document is a list of the chief customers or users of the product, usually shown as the names of other departments—or the end customer, or a supplier. Then the measures of performance are listed. There must be at least one example of each of the three types of measures. There can be more than one of each type, but this is not usually needed.

Note that these measures, unlike the first draft of business measures, are given a dimension. They must have a number by which they can be judged. They are really goals for performance and can be defined in one step, since you already have experience with the process of setting goals.

PURCHASING

Mission: To buy the materials needed for the production process

Product: Purchased production materials

Customer: Manufacturing, Assembly

Measures: Less than 0.3 percent scrap in Manufacturing due to quality defects of purchased material

Less than 0.1 percent scrap in Assembly or in the first year of product use

Less than one per thousand material deliveries delayed for lack of receipt of material

Annual increase in cost of received material less than inflation (as measured by selected high-cost items)

Figure 9–4. Purchasing measures.

ANALYTICAL METHOD

The analytical work should be done within the department itself. Hopefully, it is done by discussion with customers. It might need the help of team members and possibly the help of a systems analyst. The steps in the analytical method are listed in Figure 9–8. The first step is to list the product or products of the department. The products should already have been identified in the statement of the mission of the department. If that step was not done properly, the failure should now be apparent, and the mission statement should be reworked to tell a more accurate story.

The principal customer or customers should now be listed. It is important to list the different customers when they have different uses for the product. This too is a potential cause of conflict, since

PRODUCTION SCHEDULING

Mission: To set the rate of production of final product

Product: Production schedules

Customer: Manufacturing, Assembly

Measures: Production capacity available to satisfy 99 percent of demand

Finished product available on time to meet 95 percent of customer-requested ship dates

Total cost of unused capacity and of maintaining finished product inventory less than 5 percent of total product cost

Figure 9–5. Production Scheduling measures.

ORDER ENTRY

Mission: To balance customer demand with final product availability

Product: Open order file, Shipping Orders

Customer: Production Scheduling, Shipping

Measures: Less than 2 percent of customer orders canceled due to lack of ability to deliver product

Approved orders placed in open order file within 24 hours of receipt from customer

Operating cost of department less than 0.3 percent of total product cost

Figure 9–6. Order Entry measures.

ASSEMBLY

Mission: To assemble final product from components and subassemblies

Product: Final product (washers and driers)

Customer: Department Store Chains

Measures: Less than 0.1 percent scrap

Less than 0.1 percent failure in first year of use due to assembly error

At least 97 percent performance in meeting scheduled production rates

Assembly cost less than 7 percent of total product cost

Figure 9–7. Assembly measures.

1. Identify major product(s)
2. Identify major customer(s)
3. Ask the customers for measurements
4. Consider relevant criteria
5. Discuss reasonability
6. What if you had to pay?
7. Integrate conflicting demands
8. Note conflicts

Figure 9–8. Department measures analytical method.

the customers are the ones who should have a major say in the selection of measures of performance. All the customers of a single product do not have common requirements. Do not try to list all possible products and customers. Some of them are internal to the department and will be identified as you look at the subfunctions of each department at a later step in FA. These internal customers and their needs will also play a very important role. There is often a tendency to serve the internal customer better than the external one. In spite of all this, there is usually only one main product and one main customer. Do not worry about making the process overly complicated.

Now that you know who the customer is, ask the customer for ideas on how to measure the product. Look for criteria that relate to his or her need—the use to which the customer puts the product. Apply dimensions to the measures at this stage. There is no need to wait until a later stage, as you did for the business measurements.

When you set the dimensions on cost, the chain of command above the department head will want to have input to the process. They will be concerned about what they can afford to invest to meet the criteria for quality and timeliness. What is the *value* of the product to the user or to the business as a whole? This perception of value is a very significant factor when it comes to budgeting funds. People tend to have unrealistic expectations if they do not pay for what they receive. So it is useful to ask, "What if you had to pay for it?" This is much like setting a price on a product that is to be offered for sale by the business.

It is the task of the department representatives on the FA team to integrate any conflicting demands and to set what seem to be proper measurements. This becomes a bit more challenging when a single product is used by more than one department. For example, Accounting might want sales reports weekly for cash management, whereas Production Scheduling says that it needs sales reports daily in order to plan near-term production rates. In this case, Production Planning is likely to be making a more stringent demand for timeliness and is likely to be willing to "pay" more for the information than Accounting is.

The measures, with the dimensions, are the basis for the department strategies that you will develop two steps from now. The user will let you know later in the analysis if you have not done the job well. The strategies for each department will set the level of performance that can be accomplished. This will not in all cases be acceptable to the customer or to the one who rules on the departmental budget. This then becomes a basis for conflict and is the raw material from which to seek solutions, that is, improvements in the systems by which the business runs. Make note of any conflicts that become apparent at this step. They will be analyzed later, in Chapter 12. If you have any ideas on how to resolve a conflict, make note of them also. But do not try to obtain agreement at this stage.

IMPACT ON STRATEGIES

The measure of timeliness for purchased material will have a lot to say about what strategy Purchasing selects to source the material. It will also have an impact on Quality Control for when and how to ensure the quality of material, and on Materials Handling for where and how to store material. As an example, if material is to arrive days or weeks in advance of use, quality can be checked on receipt. But if the material is to arrive JIT, the quality must be verified before it leaves the supplier, really at the same time it is produced. JIT production leaves no time for inspecting the quality of the product at the point of use.

The measure of the amount of final product to be on hand for customer orders will have a big impact on strategies for inventorying the final product and possibly on the method of assembly. It seems that the degree of JIT production sets the pace for the degree of CIM that is needed.

MULTIPLE PRODUCTS

There are times when a single department is the source of more than one product. An example is shown in Figure 9–6 where Order Entry has to balance the needs of two different customers. It must produce an order confirmation for the end customer and an internal shipping order for the delivery of final product.

If a department supplies a number of products that are needed by unrelated units of the business, then it might be better to define multiple "departments." There is nothing rigid about how the current organization of the business is to be represented in FA. In fact, too close an adherence to the current structure of the organization can impede FA. In the case of the Order Entry example, the two products are the result of a single function. For that reason, there is no real advantage to be gained from trying to split the department.

TYPES OF PRODUCTS

In the examples shown in Figures 9–2 through 9–7, each of the three possible types of product is represented. The product of Manufacturing in Figure 9–2 is material, major subassemblies of the final product. The product of Materials Handling in Figure 9–3 is service, moving material to the point of need. Even though Materials Handling picks up the major subassemblies from Manufacturing, it is not a customer but merely a delivery agent. The product of Production Scheduling in Figure 9–5 is information, the pace at which production will take place.

MULTIPLE CUSTOMERS

The Manufacturing Department in Figure 9–2 has two customers who are likely to have different needs in terms of timeliness, if not in terms of quality and cost per unit. Materials Handling in Figure 9–3 might find itself faced with different demands from Manufacturing and from Assembly for when to move materials to the point of use. In fact, the choice of strategies might make it unnecessary for Materials Handling to move subassemblies from Manufacturing to Assembly. It is possible to design the production process so that the material moves by conveyor or by robot. The demand on Materials Handling is set by the strategies of the units it serves. Just how JIT is the production process? Just how automated is it?

MULTIPLE MEASUREMENTS

The Purchasing Department in Figure 9–4 has two measurements of quality. The impact of defective materials is different in Manufacturing and Assembly. The cost and effort to react to unusable material are different in the two departments. If a single, high measurement for quality were to apply across the board, the cost of purchased materials for Manufacturing might be higher than needed. If a single, low measurement were to apply across the board, the cost of assembly might be too high.

There are many such cases where the product of a unit should be looked at as several products, since it serves different users with different needs.

CONFLICTS

It is possible to find a number of sources of conflict in the examples. The very nature of the Order Entry Department in Figure 9–6 puts it in a contention position, since its day-to-day decisions have a direct impact on other departments and on the end customer. What serves one can be a disservice to the other.

The cost measurement of Materials Handling in Figure 9–3 could force it to operate in a manner that would prevent it from being able to meet the other measurements. The cost measure could make it fail to serve the needs of its customers by taking shortcuts or by batching materials for delivery.

The cost measurement of Purchasing in Figure 9–4 was put in place for a very good reason: If the measurement were a percentage of the cost of product, it could cause Purchasing to resist new, more expensive materials that could reduce manufacturing or assembly costs. A new, precision casting could reduce the cost of machining, because there would be less material to remove. But it would cost more, thus moving cost from Manufacturing to Purchasing.

Review the examples to see if you can find other potential causes of conflict between units.

CAVEATS

The measurements can cause shortsighted judgments if they are ill chosen. An example is the reluctance to buy better material that could reduce manufacturing cost. If at all possible, the choice of measures should promote cooperation. Measurements are guides. They will change over time when making long-term decisions due to changes in environment or technology. Any change in a business strategy will cause a change in some department strategies. Do not look at the measures as if they are carved in stone.

COMMON ERRORS

One common error is to set measures that are not practical to monitor. A measure that costs too much to obtain is self-defeating. Another common error is to select measures that place emphasis on change from the previous year. An example would be to reduce the cost of moving material by 5 percent each year. This is a sign of a lack of imagination in setting measures and would surely cause action that would transfer cost from Materials Handling to other departments. Another error is to look too hard for a single measure for any of the three criteria. This can lead to poor strategies if there are multiple products or multiple customers.

On the other hand, there can be too many measures. If multiple measures are not needed, then too many of them can be the cause of mixed signals to the producing unit.

There is always the danger of forgetting the user's point of view. It is more important than the viewpoint of the supplier. Remember, if there is no user, then there is no need for the product and no need for the unit that produces it.

CAUSES AND CURES

It is wise not to confuse the definitions of *mission, strategy, measure,* and *goal*. The sequence of steps in FA is set up for the purpose of avoiding this problem. It is aimed at helping to avoid the possibility of defining strategy before knowing the requirements to be met. The requirements are set by the measures.

A second cause of poor measures is to forget the viewpoint of the customer or to presume that you already know it. The best way to be sure that you know what the customer needs is to talk with him or her. Better yet, listen to the customer.

Shallow thinking is dangerous at any time. Ask why, why, why? Why this measurement and not some other one? Shallow thinking can lead one to forget to verify that the measurement is practical. Take a bit of time to figure out *how* the measurement might be made. For example, it is more cost-effective to keep unit sales records of automobiles than it is to do so for washing machines. The value of the product has a lot to do with the amount of money you can afford to spend in measuring it.

QUALITY CHECKLIST

A quality checklist is shown in Figure 9–9. The products must be identified. If you do not know what the product is, you do not know what to measure or how to do so. The primary customers must be identified. If you do not know who the customer is, then how can you ask the customer what he or she needs?

There must be at least one of each of the three types of measurements. They are quality, timeliness, and cost per unit. There should

1. Product(s) identified
2. Primary customer(s) identified
3. At least one of each of three measurements
   - quality
   - timeliness
   - cost per unit
4. No more than two of any measurement
5. Relevant measurement criteria
6. Measurable and affordable
7. Dimensions included

Figure 9–9. Department measures quality checklist.

be no more than two of each type of measurement. If you find need for more measures, then you should look to see if the "department"is the source of too many products. This is the time to look at subdividing the department into smaller units, at least for the purpose of FA.

Make sure that the measures are relevant and will tend to affect the performance of the mission to the best interests of the user and the business as a whole. Are the measurements practical ones that can be accomplished to some degree of accuracy without undue expense? Does each measure have a realistic dimension, a number, that will affect the choice of operational strategies for the unit?

SUMMARY

The measures of mission performance are the keys to how the day-to-day missions of the departments will be managed. Both the missions and the measures are enduring, but they still can change over time. They describe the very nature of the business and how it will operate.

The measures will have a direct impact on the choice of operating strategies for the departments. The degree of need for JIT—or of Computer Aided Engineering (CAE) or Computer Aided Manufacturing (CAM)—will be determined by the measurements and their dimensions.

The degree to which the business uses these computer technologies should not be determined just by the availability of technology, nor should it be set by industry trends. Technology should be used where it helps the business to meet its goals, to change its mission, or to reduce the cost of operation.

10
Department Subfunctions

INTRODUCTION

As stated in the last chapter, a department can be the source of a number of products. Some of them are more important than others, and some might not be needed. This will become apparent in future steps of functional analysis (FA).

Many of the products of a department are internal to the unit. They never pass out of the unit but serve the needs of the department itself. Some of these internal products might be vital to the mission of the department.

In this portion of FA, the task is to list the subfunctions of each department. A department is not a single unit with all the people doing the same thing. A typical department is divided into smaller units, subfunctions. The people in these subfunctions do different things to help the department fulfill its mission. The products, internal and external, are the keys to the breakdown of the unit into its components.

This step of FA is done within each department. It is best done through the leadership of the department's representative on the FA team. There is little need to talk to other departments in order to do this task and, for this reason, it can be done rather quickly.

DICTIONARY DEFINITION

A *function* is defined by *Webster* as "the or proper activity of a person, institution or thing."

WORKING DEFINITION

The normal or proper activity of a part of a department is to produce a product. So a *subfunction* can be defined as a portion of a department that produces a product. Most of the products of subfunctions are used within the department as aids to help it produce the product that it provides to others. As before, the product can be either material, information, or service. The products define the subfunctions: "By their works shall ye know them."

SUBFUNCTION LABEL

Each subfunction must be identified by a name or label. The component parts of the label of a subfunction are listed in Figure 10–1. A label consists of a verb and a noun: MAKE GEARS. ASSEMBLE RADIOS. REPORT SALES. DELIVER PARTS. An adjective can be placed before the noun if it is needed to distinguish the product from other things: MAKE DRIVE GEARS. ASSEMBLE FM RADIOS. REPORT DAILY SALES. DELIVER SPARE PARTS.

EXAMPLES

A set of examples are shown in Figures 10–2 through 10–7. They are based on the same six departments that have been used in previous chapters. The subfunctions are meant to be examples of how each department might be subdivided for the purposes of FA. The breakdown of a department will vary a great deal from company to company. The possible subfunctions of, Manufacturing, Materials Handling, Purchasing, Production Scheduling, Order Entry, and Assembly are shown in Figures 10–2, 10–3, 10–4, 10–5, 10–6, and 10–7, respectively.

METHOD OF REPRESENTATION

The list of subfunctions is simply added to the document that describes the mission of each department. You will finish this docu-

1. Verb
2. (Adjective)
3. Noun

Figure 10–1. Subfunction label.

MANUFACTURING

Mission: To produce major components of Hydroclean products

Product: Major subassemblies

Customer: Assembly, Field Service

Measures: Less than 0.3 percent of production scrapped in production, at assembly, or in the first year of use

Less than 1 percent of assembly delay per week for lack of manufactured components

Cost per unit 10 percent less than delivered cost of purchased alternative

Subfunctions: Schedule Manufacture
Produce Components
Produce Assemblies

Figure 10–2. Manufacturing subfunctions.

ment two chapters from now when you describe the operating strategies for each unit.

ANALYTICAL METHOD

The steps in the analytical method are listed in Figure 10–8. When you first stated the mission of each department and listed its products, you probably found it hard to limit the list to one product or, at most, two. Now is the moment when you should list the other products that came to mind at that time. Some of them might be minor products for use by other departments. Some are only used internally within the unit itself to assist in production of the main product of the unit. For example, a bin catalog is a product of

MATERIALS HANDLING

Mission: To move materials to the point of need

Product: Delivered production materials

Customer: Manufacturing, Assembly

Measures: Less than one error per thousand deliveries, including misdeliveries and damaged materials

Less than 1 percent delay in weekly production time in Manufacturing or Assembly due to late delivery of available material

Material handling cost less than 3 percent of total production cost of product

Subfunctions: Plan Staffing
Receive Material
Store Material
Deliver Material

Figure 10–3. Materials handling subfunctions.

Materials Handling to help keep track of where parts are stored. It never leaves the unit but is vital to the function of Materials Handling.

Also make a list of the subdivisions of the department as shown on the organization chart of the unit. Ask what the products of these subdivisions are. Between these two lists, one from a product view and one from an organization view, you should have all the raw material from which to list subfunctions. You should end up with from three to six of them. If your two lists have less than three significant products, both internal and external, chances are that the department is best described as a part of another department. If you

PURCHASING

Mission: To buy the materials needed for the production process

Product: Purchased production materials

Customer: Manufacturing, Assembly

Measures: Less than 0.3 percent scrap in Manufacturing due to quality defects of purchased material

Less than 0.1 percent scrap in Assembly or in the first year of product use

Less than one per thousand material deliveries delayed for lack of receipt of material

Annual increase in cost of received material less than inflation (as measured by selected high-cost items)

Subfunctions: Select Sources
Establish Contracts
Release Orders
Monitor Performance

Figure 10–4. Purchasing subfunctions.

identify more than six significant products, look for some basis on which to separate the original department into two separate "departments." Do not worry that you are going contrary to the traditional organization chart. You might have found a reason why the organization itself should change.

Remember to label the subfunctions by verb, adjective (if needed), and noun. In simple terms, the label says "MAKE SPECIFIC PRODUCT."

PRODUCTION SCHEDULING

Mission: To set the rate of production of final product

Product: Production schedules

Customer: Manufacturing, Assembly

Measures: Production capacity available to satisfy 99 percent of demand

Finished product available on time to meet 95 percent of customer-requested ship dates

Total cost of unused capacity and of maintaining finished product inventory less than 5 percent of total product cost

Subfunctions: Track Finished Inventory
Schedule Final Products
Plan Material Requirements

Figure 10–5. Production scheduling subfunctions.

TIMING

Timing is an important factor to consider in this step of FA. Some activities occur daily, as in assembly of final product; some activities occur in calendar cycles, as in the preparation of budgets; some occur infrequently, as in the design of a new product for sale; and some occur minute by minute, as in the JIT delivery of parts for assembly.

What might appear to be a single product can at times be two products. As an example, think about the maintenance of equipment. This is a service that is performed on a timed schedule to prevent failure. It is also performed on demand when the equipment breaks down. The two subfunctions might be PERFORM PREVENTIVE MAINTENANCE and REPAIR FAILED EQUIPMENT.

ORDER ENTRY

Mission: To balance customer demand with final product availability

Product: Open order file, Shipping Orders

Customer: Production Scheduling, Shipping

Measures: Less than 2 percent of customer orders canceled due to lack of ability to deliver product

Approved orders placed in open order file within 24 hours of receipt from customer

Operating cost of department less than 0.3 percent of total product cost

Subfunctions: Verify Credit
Set Delivery Date
Issue Shipping Orders
Maintain Order File

Figure 10–6. Order entry subfunctions.

COMMON ERRORS

A common error is to be bound by the existing organization chart. An argument can be made to place a given subfunction into any one of several departments. For example, any one of a number of departments could plan material requirements—Assembly, Manufacturing, Production Scheduling, Purchasing, or Materials Handling. Try to place the subfunctions under departments where the flow of products is logical.

A related error is to "bog down" in conflict between two or more departments over who is the owner of a subfunction. It is not important to settle who has what "turf" but to determine what tasks are to be performed. If such a conflict occurs, list the topics of conflict and the reasons for them. Then move on, since you will have time to resolve the conflict later. One way of reducing conflict

ASSEMBLY

Mission: To assemble final product from components and subassemblies

Product: Final product (washers and driers)

Customer: Department Store Chains

Measures: Less than 0.1 percent scrap

Less than 0.1 percent failure in first year of use due to assembly error

At least 97 percent performance in meeting scheduled production rates

Assembly cost less than 7 percent of total product cost

Subfunctions: Schedule Assembly
Assign Assembly Tasks
Assemble Product
Test Product

Figure 10–7. Assembly subfunctions.

1. Identify other external products

2. Identify internal products

3. List next level of organization

4. Define departments of three to six subfunctions

Figure 10–8. Subfunction analytical method.

is to name the departments with labels that are different from those on the organization chart. Create a logical organization to represent what happens in the physical organization.

When the charter of the FA study is based on a desire to change the mission or strategies of the business, it is possible to confuse past and future subfunctions. Make sure that your time frame is the future, especially if you know what you want to change.

Does the title of a subfunction contain the label of a product? If it does not, then you most likely have not identified a valid subfunction. Like a business or a department, the mission of a subfunction is to provide a product. The difference is that the subfunction need not be a specific, identifiable part of the organization.

CAUSES AND CURES

It is hard to avoid being bound by the existing structure of the organization. Use it only to identify products and then label the subfunctions that provide the products. These subfunctions could be part-time responsibilities of people who also perform other subfunctions.

Do not let the team get caught up in organizational conflict as to the ownership of a subfunction. Make a note of it and move on. You will have plenty of time to work out solutions to such conflict two chapters from now. In fact, you should not argue between departments as you do this step. Each team member should do the analysis for his or her own department. Of course, the team member should seek the help of others in the department. The overlaps and omissions in responsibility will become apparent in the next steps of FA as you read what the various departments have said in their documentation.

If you find that you are confusing past department activities and future activities, you probably want to stick to future activities. After all, you are doing this exercise to define systems for the future.

QUALITY CHECKLIST

A checklist is shown in Figure 10–9 to help you ensure the quality of your work in this step of FA. There should be no less than three and no more than six subfunctions in each department. Is the title

1. Three to six subfunctions
2. Verb, (adjective), noun
3. No *and*'s
4. Is the noun a product?

Figure 10–9. Subfunction quality checklist.

restricted to verb, adjective, and noun? This seems to be an unnecessarily rigid rule. It is not. The discipline forces you to be sure that you have indeed identified a subfunction that provides a product. It is the product that is most important. There should not be any need to use the word *and* in any subfunction title. If you are tempted to do so, you are dealing with two separate subfunctions, and it is best to just identify them as two and move on.

Does the noun identify a true product? Is the adjective necessary to differentiate it from other products that are similar? As an example, the many subfunctions of a business might include such labels as BUY FINISHED PARTS, MAKE PIECE PARTS, REWORK SCRAP PARTS, DELIVER SPARE PARTS. Parts come in all sorts of flavors.

SUMMARY

This step of FA is deceptively simple and might be performed in an hour or two for any given department. But it is a very important step, none the less. It is the basis for the next step of FA in which you will begin to define operating strategies for each department. The internal products of a department imply strategies. Otherwise, why do they exist?

The need to gather some subfunctions under new departmental organizations might become apparent later in the analysis as you work to resolve the conflicts.

11
Department Strategies

INTRODUCTION

This is the final step in functional analysis (FA) at the departmental level. The next steps are at lower levels. They are concerned with how each subfunction works but will have an impact on the departments and will most likely cause you to change some department strategies and measures of performance.

It was important for you to identify the subfunctions of each department as a necessary step to the definition of department strategies. Most such strategies are not explicitly documented but are implicitly understood by those whose work is guided by them. They are often embedded in how the department is organized and operates and in what products it produces for itself and for others.

This is the last step of the process (Chapters 8 through 11) of developing a description of each department in the FA project.

WORKING DEFINITION

The definition of a *department strategy* is like the definition of a *business strategy,* since they serve the same ends at different levels of the organization. A *department strategy* is a statement of a principle of operational behavior or judgment that serves as a guide for day-to-day decisions. These guides are needed so that the department measures of accomplishment of mission will be acceptable to management and will support the needs of the business. The systems of a department are based on these strategies.

As stated in Chapter 7, a strategy should be supportable with a reason. There must be a reason why the chosen strategy is preferable to at least one alternative strategy. The alternate strategy must not be trivial but must make good business sense.

STRATEGIC TOPICS

The list from Chapter 7 of strategic topics for a business has been reproduced here in Figure 11–1 as a reminder. The same set of

1. Pricing
2. Differentiation
3. Segmentation
4. Distribution
5. Financing
6. Procurement
7. Manufacturing
8. Finished Inventory
9. Product Support
10. Product Change
11. Technology
12. Quality

Figure 11–1. Strategic topics.

topics are possible subjects of strategies for a department. But of course, this time the topics will be from a departmental point of view.

Because of the rapid growth in computer technology, you want to consider the possibility of strategies based on JIT, factory automation, CAE, CIM, or any such "hot topic." Ask yourself if any of these technologies are needed or if they might bring significant value. The answer to this question might be easier after Chapter 16. Then you will test the need for computer aid to implement the strategies as they are reflected in systems.

EXAMPLES

The examples used in the previous chapters have been retained and supplemented in Figures 11–2 through 11–7. The strategies have been added as the final piece of documentation of the departments. Two or three strategies are shown for each department: Manufac-

MANUFACTURING

Mission: To produce major components of Hydroclean products

Product: Major subassemblies

Customer: Assembly, Field Service

Measures: Less than 0.3 percent of production scrapped in production, at assembly, or in the first year of use

Less than 1 percent of assembly delay per week for lack of manufactured components

Cost per unit 10 percent less than delivered cost of purchased alternative

Subfunctions: Schedule Manufacture
Produce Components
Produce Assemblies

Strategies:
1. The manufacture of major subassemblies will be driven by the production schedule on a JIT basis.
2. Components of major subassemblies will be manufactured only if it can be done at 10 percent less cost than outside purchase.

Figure 11–2. Manufacturing strategies.

MATERIALS HANDLING

Mission: To move materials to the point of need

Product: Delivered production materials

Customer: Manufacturing, Assembly

Measures: Less than one error per thousand deliveries, including misdeliveries and damaged materials

Less than 1 percent delay in weekly production time in Manufacturing or Assembly due to late delivery of available material

Material handling cost less than 3 percent of total production cost of product

Subfunctions: Plan Staffing
Receive Material
Store Material
Plan Daily Requirements
Deliver Material

Strategies:
1. Maintain a central inventory of production material equivalent to a week of use.
2. Deliver material overnight in accord with a breakdown of the production schedule.
3. Provide hourly pickup of final product from Assembly.

Figure 11–3. Materials handling strategies.

PURCHASING

Mission: To buy the materials needed for the production process

Product: Purchased production materials

Customer: Manufacturing, Assembly

Measures: Less than 0.3 percent scrap in Manufacturing due to quality defects of purchased material

Less than 0.1 percent scrap in Assembly or in the first year of product use

Less than one per thousand material deliveries delayed for lack of receipt of material

Annual increase in cost of received material less than inflation (as measured by selected high-cost items)

Subfunctions: Select Sources
Establish Contracts
Release Orders
Monitor Performance

Strategies:
1. Contract only with suppliers who have in-house quality programs to meet our requirements.
2. Contract for daily delivery to a schedule established a month in advance.
3. Rely on Assembly, Manufacturing, and Materials Handling reports for supplier performance monitoring.

Figure 11–4. Purchasing strategies.

PRODUCTION SCHEDULING

Mission: To set the rate of production of final product

Product: Production schedules

Customer: Manufacturing, Assembly

Measures: Production capacity available to satisfy 99 percent of demand

Finished product available on time to meet 95 percent of customer-requested ship dates

Total cost of unused capacity and of maintaining finished product inventory less than 5 percent of total product cost

Subfunctions: Track Finished Inventory
Schedule Final Products
Plan Material Requirements

Strategies:
1. Plan the assembly schedule weekly to a horizon of four weeks based on open orders.
2. Plan material requirements weekly based on the assembly schedule and a forecast of service parts demand.

Figure 11–5. Production scheduling strategies.

turing, Figure 11–2; Materials Handling, Figure 11–3; Purchasing, 11–4; Production Scheduling, Figure 11–5; Order Entry, Figure 11–6; and Assembly, Figure 11–7.

The statements are numbered to make it easy to find the supporting rationale, as was done for business strategies. Refer to the

ORDER ENTRY

Mission: To balance customer demand with final product availability

Product: Open order file, Shipping Orders

Customer: Production Scheduling, Shipping

Measures: Less than 2 percent of customer orders canceled due to lack of ability to deliver product

Approved orders placed in open order file within 24 hours of receipt from customer

Operating cost of department less than 0.3 percent of total product cost

Subfunctions: Verify Credit
Set Delivery Date
Issue Shipping Orders
Maintain Order File

Strategies:
1. Fill a percentage of a customer order if finished inventory is in short supply.
2. Set promise dates for remaining units on a first-come, first-serve basis.

Figure 11–6. Order entry strategies.

example in Figure 11–8. The supporting rationale shows why the chosen strategy is preferable to alternatives. The supporting rationale is not as vital as is the one for business strategies, since distribution of these strategies is not as wide. But it is still of value within the department, and it will help in resolving conflicts with other units. An example of a rationale for the first Manufacturing strategy is shown in Figure 11–8.

ASSEMBLY

Mission: To assemble final product from components and subassemblies

Product: Final product (washers and driers)

Customer: Department Store Chains

Measures: Less than 0.1 percent scrap

Less than 0.1 percent failure in first year of use due to assembly error

At least 97 percent performance in meeting scheduled production rates

Assembly cost less than 7 percent of total product cost

Subfunctions: Schedule Assembly
Assign Assembly Tasks
Assemble Product
Test Product

Strategies: 1. Assemble product in daily batches by type (washer versus drier).

2. Vary assembly task assignments each week.

Figure 11–7. Assembly strategies.

ANALYTICAL METHOD

Figure 11–9 contains a list of things that you should consider when defining strategies for your department. Look for the current set of departmental strategies. They should be more or less on target, or else the business is in a lot of trouble. But do not be bound by these strategies. Consider the measures you have defined and ask yourself if they might not make it necessary to change some strategies.

MANUFACTURING STRATEGY RATIONALE

1. The manufacture of major subassemblies will be driven by the production schedule on a JIT basis.

   Major subassemblies take up a large amount of space. The costs of storage and handling of subassemblies can be minimized if the production is tied directly to the schedule for assembly of final product, with material flowing directly from the subassembly process to the assembly line.

   With this direct flow, any quality defects in the subassemblies are more likely to be found and corrected.

   This JIT manufacture requires a material planning and procurement system that is also JIT. The possible disruption of the assembly process due to failure in the procurement or manufacturing processes is considered a necessary risk and a motivator toward prompt resolution of quality and delivery problems.

Figure 11–8. Manufacturing strategy rationale.

Ask why the internal products are needed. Do they really make it easier or less costly to produce those products that are needed outside the department? Review the cycle time and response time asked of externally used products. What are the quality requirements of your customers? Can you meet them? Are the requirements excessive? Do your strategies make it possible for you to meet the needs of your customers?

Now that you have looked at the needs of your customers, ask about the degree to which your department's needs are being met by the products it receives from others. Ask what reliability you can expect of the products used by your department, which includes

1. Consider existing strategies
2. Evaluate need for internal products
3. Ask demands on externally used products
4. Ask demands on externally provided products
5. Examine the impact of physical constraints
6. Evaluate lead-time requirements
7. How does the department order products?
8. How does the customer order products?

Figure 11–9. Department strategy analytical method.

things such as forecasts, orders, and supplies. Do they meet your needs for quality and timeliness? Does the supplying department agree with your desired level of timeliness and quality? Are they adequate for you to accomplish your mission?

Consider the physical limitations in which you work, such as warehouse space. Do they constrain your use and storage of materials and equipment? Ask what lead time is needed, such as for raw material from suppliers. Ask how your suppliers know what you need and when you need it. Should you try to set up a better way to tell a supplier when to deliver to you? Ask how you might best know when to deliver product to those whom you serve. Should they tell you delivery by delivery, or should they provide you with a long-term schedule?

The above topics need not be considered in any special sequence. They all are important. The list is not meant to be all-inclusive but is intended to serve as a guide to the types of things that you should consider when setting strategies.

COMMON ERRORS

A common error is to develop too many strategies. There is a tendency to try to cover all possible situations. A few significant strategies are a better guide to daily work than a book full of writings that are never read.

A second common error is to focus on the internal convenience of the department only. It is not to the benefit of the business as a whole if you maximize the performance of the unit at the expense of your customer.

A third common error is to phrase strategies that say nothing other than that which is obvious. They are "motherhood and apple pie" strategies that bring disrespect to those other strategies that have real value.

CAUSES AND CURES

One cause of poor strategies is a quick and superficial analysis of the need for strategies, perhaps with too much willingness to accept the strategies that are already in place. Or it might be that the team member already has in mind solutions for problems that have yet to be defined. The way to avoid this trap is to perform this FA step with care. Look at each product, both internal and external; ask if it really serves a valid need; ask if its measures are correct.

JIT , CIM, CAE, and other modern techniques bring great value to industry. But it is not unusual for a business to grasp at one of these new methods without first asking if there is commensurate value to be obtained. Do not use these tools because it is the popular thing to do. Do it if you have gone through an FA and have found a reason for the change.

Another cause of poor strategies is the narrow viewpoint, the person who thinks only of his or her own good or department's good. Take the time to consider each customer and each supplier. Are their needs being met by your strategies? If not, and if there is not an evident solution, take note of the conflict for later resolution.

It is possible to try to be too detailed in thinking through the strategies. If things get to be confusing, take note of the subject as a conflict and hold it for analysis at a later stage of FA.

QUALITY CHECKLIST

A checklist for assessing the quality of your strategies is shown in Figure 11–10. There should be from two to five strategies for each department. The number of strategies should be no more than the number of subfunctions. Normally, there is one strategy that is

1. From two to five strategy statements
2. No more strategies than subfunctions
3. Rationale for each strategy
4. Identified potential conflict

Figure 11–10. Department strategy quality checklist.

aimed at each subfunction or at several of them. Since there should be from three to six subfunctions, it is only reasonable that the number of strategies should be the same or a bit less. You might have reason for very many or very few; but if you do, you have probably fallen into one of the errors.

Each strategy must be accompanied by a clear rationale, which should include a comparison with alternatives, implicitly or explicitly telling why the chosen strategy is to be preferred to them.

It is not necessary, or even to be expected, that a conflict will result from each strategy. But if conflict has been identified, it is a good sign that the strategies are the result of creative thinking and that FA is beginning to give you value in return for your effort. Conflict is only bad if it is ignored or unmanaged.

CONFLICTS

The sources of a number of potential or real conflicts are illustrated in the strategies shown in Figures 11–2 through 11–7. The second strategy of Materials Handling (Figure 11–3) is in conflict with the first strategy of Manufacturing (Figure 11–2). To see this, refer to the rationale for the Manufacturing strategy in Figure 11–8.

The third strategy of Materials Handling (Figure 11–3) is the source of a potential conflict with Assembly (Figure 11–7). There is a conflict if adequate storage of finished product is not provided at the assembly area. It is possible that there is not adequate floor space in some areas to hold an hour's output of production.

The second strategy of Purchasing (Figure 11–4) could be in conflict with Production Scheduling (Figure 11–5) if the latter changes the assembly schedule within the four-week horizon. This could cause Purchasing the need to renegotiate commitments from suppliers.

The third strategy of Purchasing (Figure 11–4) can be in conflict with the Quality (not included in examples) organization if there is a perceived overlap of responsibility. The Quality Department could say that it has the authority to monitor supplier performance.

The first strategy of Production (in examples) Scheduling (Figure 11–5) can be in conflict with Marketing (not included) if the planned response time to customer orders is less than four weeks. If Marketing cannot maintain an order board of at least four weeks, Production Scheduling will not have the information it needs in order to set a reliable assembly schedule.

The same first strategy of Production Scheduling (Figure 11–5) can be in conflict with the first subfunction of Assembly (Figure 11–7). Both units seem to think that they are the ones responsible for setting the assembly schedule. This is a territorial dispute.

The Order Entry (Figure 11–6) strategies can potentially conflict with the Sales Department (not included in examples), which will probably want to give special attention and service to large or hard-to-please customers.

If the root causes of these potential conflicts are not apparent at this time, the discussion of conflict in the next chapter should clear up any confusion.

SUMMARY

The strategies should be a result of an in-depth analysis of products and customers, of business strategies, and of department measures and goals. There should be an exchange of ideas among the departments as the strategies are developed. "Can you live with this?" "Can you provide me with that?"

Unless communication and cooperation have been exemplary in the past, this step of FA should result in a number of new department strategies. At the same time, you should be able to identify a number of points of conflict among departments.

Remember not to try to resolve conflict too early. Several more steps of FA are needed before this can be done with assurance. The intent of FA is to lead you in a logical path to improved systems. Do not rush it. The result of this step of the analysis is the completion of a document describing each department involved in the FA.

12
Organizational Conflicts

INTRODUCTION

Conflict is a natural state of affairs among many units within a business. This conflict stems directly from the business strategies and the measures of performance of the units as they strive to do what they feel is expected of them. Conflict can be very destructive to the business as a whole if it is not managed. But if it can be identified—and it can be through the use of Functional Analysis (FA)—then its causes can also be identified. And solutions can be found through FA teams so that the conflict can be minimized and managed.

This is a step in the FA process where team thinking is mandatory in order for you to obtain the system improvement that you need.

DICTIONARY DEFINITION

Webster says that the word *conflict* means "the interference of one personal interest with another." The noun *problem* means "a question proposed for solution." As an adjective, it means "dealing with matters arising out of conflicting social values and relationships." It would seem that a problem is a barrier to progress that is caused by a conflict of values.

WORKING DEFINITION

For the purposes of FA, *active conflict* can be considered to be interference between units within a business. This might be a case where two departments each think that they have the same responsibility, or where one unit wants to take a responsibility away from another. But there is another, more common and less apparent type of conflict that might be called *passive conflict*. This happens when there are voids in products or services. A void occurs when nobody provides a thing that is needed. It also happens when the product lacks in quality or timeliness or when the cost exceeds the perceived value.

In simple terms, conflict is disagreement among business units over the products provided from one unit to the other. Whatever the cause or the nature of the conflict, it is a problem that needs to be solved so that the business can run as it should.

EVERYONE PRODUCES A PRODUCT

Every unit of an organization in some way or other makes a product. This point was made in the introduction to the concept of product in Chapter 2. Figure 12–1 shows a model that can be used to illustrate the concept of product and to make it more useful for FA. The product of a business unit can be material, information, or service. It is produced to serve the needs of a customer or a user. If there is no product or no user, then there is no reason for the producing business unit to exist.

Customers come in two flavors: They can be internal or captive customers, part of the business itself; or they can be external customers who are totally free to buy or to reject the product of the business.

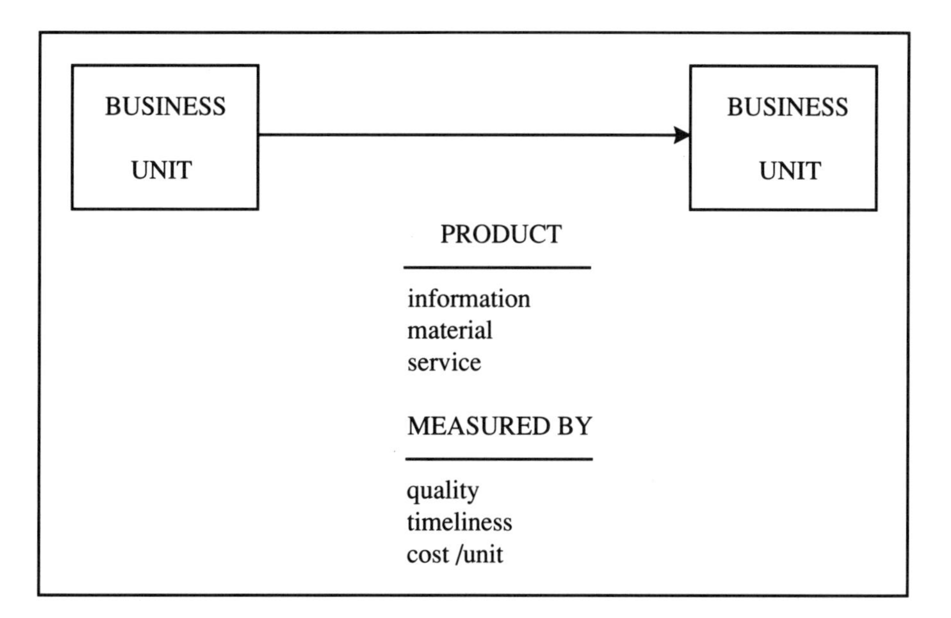

Figure 12–1. Function of a business unit.

There are three measures of how well the supplying unit provides the product for the best interests of the customer. These measures are quality, timeliness, and cost per unit. The most immediate point of conflict among business units is a difference of opinion as to what the dimensions should be on these measures. This difference of opinion is a symptom, a clue to the ultimate cause of the conflict, which is usually found in opposing missions and strategies among the units.

AREAS OF CONFLICT

Figure 12–2 shows the areas in which the unit that supplies the product and the unit that needs it can be in mutual conflict. A very significant source of conflict is the cost per unit of the product. This problem is caused by the fact that there is usually no transfer of charges between units for internal products within a company. So why should this cause conflict? It does so in two ways.

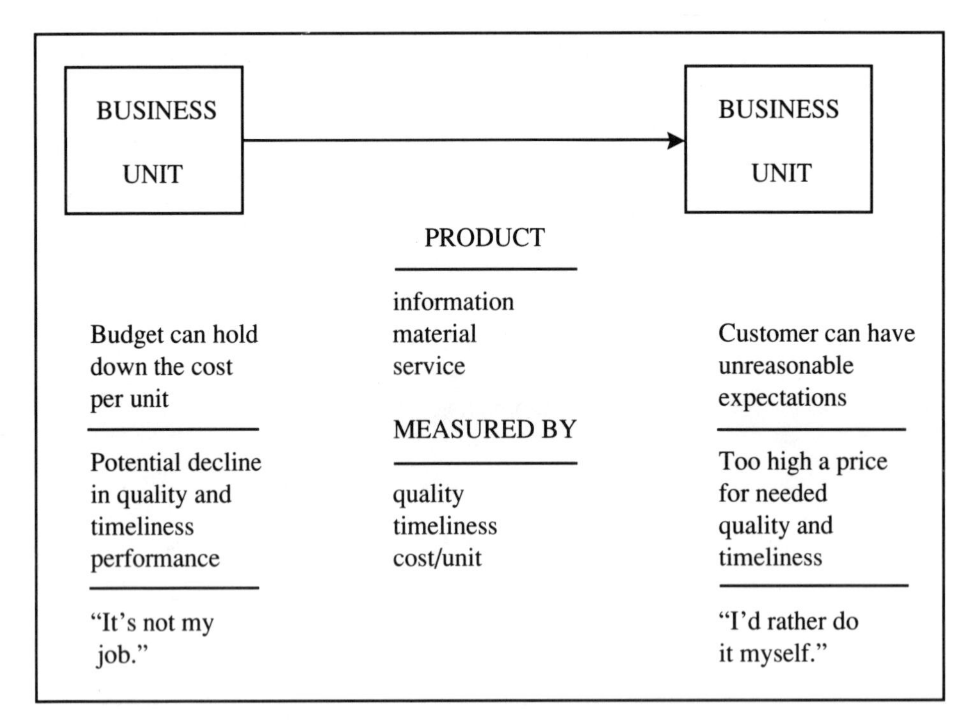

Figure 12–2. Potential areas of conflict.

It is quite common for budget constraints on the supplying unit to cause a decrease in its ability to meet the user's needs for quality and timeliness. If costs are cut, something has to suffer. Ask yourself how much influence one internal unit has over the budget of its neighbor. An internal customer who does not have to pay for the product can tend to have unrealistic expectations. He or she can ask for quality and timeliness of product in excess of what he or she really needs. The supplying unit is not always sympathetic to these needs, especially if it is obvious that the demands are excessive. A classic case is the design engineer who places very tight tolerances on the parts that he designs, since he "knows" that the factory will never make the part according to specifications. The factory then makes the part to only 60 percent of designed tolerances, since it "knows" that the design engineer is overly demanding. This is sometimes called "the games people play."

The supplier can be ignorant or uncaring of the needs of the user. The author has had much experience with the Annual Quality Improvement (AQI) program of the famous quality expert, Dr. Joe Juran. His program is aimed at rooting out and eliminating the chronic causes of scrap, loss, or waste. During the course of a large number of AQI projects, the author has never seen a case of chronic scrap where there was not also a failure in the feedback loop. The user was dissatisfied with an internal product, but the supplier did not know about it. Maybe the user had complained at one time and had then given up hope for improvement. Maybe the user had never complained at all. There are even cases where the supplier and the user did not even know about each other. The larger the business, the more likely this is to happen.

There can be an unwillingness of one unit to rely on someone else for the product it needs. This can happen because that other unit has chronically failed to deliver as needed. It can happen because the supplier never seems to be able to obtain the funds needed to deliver the product to the user's satisfaction. Or it can be a personality-driven preference: "I would rather do it myself," because I do not like the other fellow, or because I do not like to be dependent on anyone, or because I want to show my superiors how valuable I am. Or the supplier can prefer not to deliver to this specific user. Never underestimate the impact of personality on business systems.

There can be an unwillingness for the supplier to accept responsibility for a product, even though he or she might be the logical source for it. This could be due to budget constraints: "I have more important things to do." It could be because of inconvenience. "I do not want to be bothered." It could be because of personalities: "I do not like that jerk." An example could be a Purchasing Department that tells the Quality Department: "You track supplier performance for yourself. I do not have the time to do so."

The product can be in a form that is unusable to the customer. A typical example is a CAE graphics drawing of a part, where there are several users of the drawing. It is drawn in a format that is handy for the engineer. But the format makes it necessary for the tool designer and the parts programmer to recreate their own image of the part in order to do their job: "I will draw it in the format that best suits the needs of my department; you must adapt your procedures to compensate."

EXAMPLES

A number of examples of conflict among business units were listed in the previous chapter. These conflicts tend to stem from incompatible missions or strategies or measures. After all, missions lead to measures, and they in turn lead to the strategies by which the units operate. Two of these conflicts are documented here in Figures 12–3 and 12–4. The examples show how you might record the conflicts that your team identifies as you go through the process of FA.

The example in Figure 12–3 is caused by cost concerns, which cause the two departments to have incompatible ideas about how often material should be delivered to the point of need. Manufacturing wants production material to be delivered at intervals throughout the day in order to operate on a JIT basis. But Materials Handling, trying to minimize the cost of moving material, prefers to deliver a day's batch of material each night. The conflict, obviously, is a disagreement between supplier and user as to the measure for timeliness. There are a number of possible solutions to the conflict. Certainly, a workable solution must be found. One potential solution, not necessarily the best one, is shown in Figure 12–3. Like most solutions, it requires a compromise from the ideal, as seen differently by the two business units.

The example in Figure 12–4 is caused by a difference between the response time need of the end user and the planned delivery capability of the supplier. Obviously, it is not practical to order material

CONFLICT

Between: Manufacturing
Materials Handling

Subject: Delivery of production materials

Cause: Manufacturing JIT production strategy

Materials Handling strategy of overnight delivery of production materials

Solution: Storage at the manufacturing cell for a day's quantity of small materials

Hourly delivery of large materials

Figure 12–3. Resolved conflict.

CONFLICT

Between: Production Scheduling
Purchasing

Subject: Changes to the assembly schedule

Cause: Production Scheduling strategy to plan the assembly schedule to a two-week horizon

Purchasing strategy to establish a supplier delivery schedule a month in advance

Solution: Optional solutions to be developed

John Jones of Production Scheduling
Sara Smith of Purchasing

Due March 15, 1990

Figure 12–4. Unresolved conflict.

before having any idea how much is needed. The solution to this problem could involve a number of other units, including Sales and Assembly. This is an indication of the need for teamwork. Note that the solution to this conflict is not yet documented. This will most often be the case at this stage of FA. The solutions to conflicts are not often obvious.

METHOD OF REPRESENTATION

Figures 12–3 and 12–4 show a format for recording conflicts. This format has proved to be very convenient in a number of FA studies. Just like all the other documents that are produced during FA, brevity is of great benefit, so long as clarity is not sacrificed. The idea is to record the important ideas in such a fashion that they can be quickly read and comprehended. How many times have you seen a thick, bound, impressive-looking report that is never read but just gathers dust on a shelf?

The conflict document should show the names of the units that are in conflict and the subject of the conflict. It should contain an identification of the cause of the conflict. Most often this cause stems from opposing missions or measures or strategies. Be careful of the language that is used to express the conflict and its cause. It is easy to fall into the trap of using "loaded" language, which implies fault or blame. Most conflict stems from valid but opposing business goals, not from personalities or defects in human nature.

If the solution to the conflict is readily apparent, record it at the bottom of the document. This is shown in Figure 12–3. Most of the time, the solution will not be apparent when the conflict is first identified. In this case, record who is responsible to seek a solution, and a target date indicating when the solution is needed. This is shown in Figure 12–4.

Write a document to record every conflict, even if it is quickly solved. It helps the sponsor to see the value of the work of the FA team. Each conflict that you record is a conflict that had not been resolved before, possibly not even recognized before.

ANALYTICAL METHOD

Figure 12–5 shows a series of places to look in order to find conflicts. It is better to find them now rather than to find them in operating practices after the FA has been completed. Even though

1. Look for redundant products
2. Look for redundant information sources
3. Look for missing products
4. Look for disagreements on measurements
5. Look at who relies on whom
6. Iterate as necessary

Figure 12–5. Organizational conflicts analytical method.

you work hard to identify conflicts at this stage of the study, you will still find some at later stages.

Look for redundant products, where two different units are producing the same or similar products. This can happen because the units are not aware of the redundancy. Often it happens because one unit is not willing to accept the product from the other for any of the reasons stated earlier in this chapter. Look for redundant sources of information. A unit can seek the same or similar information from more than one source, because none of them is considered to be fully reliable in terms of quality or timeliness. Look for a void in products, which occurs when a unit needs a product—most often, information—but has no place from which to obtain it. There is usually a logical source, which is unwilling or unable to deliver what is needed. Sometimes the logical source is not apparent to the unit that needs the product.

Look for disagreement over product measurements. A supplying unit can be able to provide the product needed by another unit but unwilling or unable to provide it to the satisfaction of the user. Look at who is willing to rely on whom. This is where you might find personality conflict as well as mission conflict. For this reason, it is best to look gently. But keep in mind that unidentified conflict is always unresolved conflict. And unresolved conflict is detrimental to the success of the business.

Do not try to look too deeply at this time. Do not think that you must root out all conflict or that you must resolve it all at this stage of FA. There will be ample opportunity to identify more conflicts as you perform the subfunction analysis in the next few steps of FA. Figure 2–3 shows that the analysis of conflict really runs concurrently with three other steps of FA. The conflict will all be resolved in the step described in Chapter 15.

Be willing to iterate, to go back through previous steps as you seek solutions to the conflicts. Quite often the solutions lead to, or are based on, changes in missions, measures, and strategies. Do not rush to find solutions if there is disagreement or if you feel that deeper analysis should be done first. It is better to take the time to do a good job and to allow people to make the necessary mental adjustments to change. Rather than hurry the process, it is often prudent to assign responsibility and a target date for the solution to problems that need deeper thought.

Be careful when you deal with conflict that involves personalities. In this case, you should be guided by the culture of the company. Some companies are willing to deal with such conflict openly; some are not. But always look for *what* is wrong rather than *who* is wrong.

COMMON ERRORS

The first type of error during this phase of FA is shallow thinking. This happens when the team takes only a cursory look at the documentation to date or does not take the time to think through the implications of the missions, measures, and strategies of related units.

The second type of error is to fall into the trap of assigning personal blame for organizational conflict. It is well to remember that conflict is a natural state of affairs. You can lose more than you might ever hope to gain if you let personalities enter into the issues. To point a finger is to lose the ability to find solutions to conflict. Teamwork is based on goodwill.

One more serious error is trying to work out solutions before adequate analysis has been done or before the team has fully comprehended the nature of the conflict. Do not be in a hurry. The procedures that result from the FA study will be with you for a long time.

CAUSES AND CURES

There are two main causes of shallow thinking. The first cause is the person who wants to think only from a narrow, departmental bias. This person only cares about solving his or her own unit's problems and does not really care about the welfare of other units, or even of the business as a whole. The second cause is a lack of time to do the analysis properly. The only known cure is to have regular team meetings to review departmental documents as a group and to evaluate the implications of missions and strategies as a group. It helps if the leader is strong and if the sponsor is understanding about the expenditure of time.

Many conflicts simmer for years without solution. The units find ways of working around them, because they have been unable or unwilling to address them straight up. These cases usually make themselves apparent in the form of personalized hostilities. The cause of each problem is laid at the feet of one or more "guilty" parties. This type of discussion leads nowhere, except possibly to further frustration. It takes a strong team leader who can steer the discussions away from the shoals of personality.

Impatience in any form can cause conflicts to be overlooked. The same impatience can damage the quality of the results of FA at any step. It happens if the sponsor, team leader, or team members are in a hurry for results or are unwilling to spend the time needed. After all, thinking is hard work and can even be dangerous. There is the danger of learning something you do not want to know. Again, the cure takes the form of a strong sponsor who follows up regularly on progress and a strong team leader who keeps things on track.

Always remember that it is better to fix problems up front than to live in a messy house and regret the opportunity that has been lost.

QUALITY CHECKLIST

A quality checklist for the documentation of conflict is shown in Figure 12–6. All sections of the document should be filled in. If they are not, you are guilty of the impatience that was discussed above.

The causes of each point of conflict should be reasonably identified, at least the immediate causes. Often the person who writes the description of the causes says too little, leaving it up to the

1. All topics filled in
2. Causes identified
3. Terse, but descriptive, text
4. Solution included or assigned
5. Date for an assigned solution task

Figure 12–6. Organizational conflicts quality checklist.

imagination of the reader to read between the lines. Note that in the two examples not only was the cause stated but also the result. In Figure 12–3, the cause of the conflict is given as a difference of opinion as to the measure of timeliness. Figure 12–4 asserts that the causes of the conflict are specific strategies and that the results of the two strategies are two incompatible time horizons for planning.

This is the first time that you seem to have been urged not to be terse in your writing of results. This is not really the intent. It is still good to be terse, but the text must be descriptive and capable of standing alone. For each identified conflict, there should be (1) either a solution or an assigned responsibility for developing a solution and (2) a target date for when that must be done.

SUMMARY

The first set of conflicts that your team documents should illustrate the value of the FA technique. Unless the work of the team has been superficial, you should have identified some conflicts that you did not know existed. And you should also have identified old conflicts that had been taken to be the natural state of affairs. But by also pinpointing the causes of these chronic conflicts, you have taken a giant step toward finding a solution for them.

There will be many more conflicts that you will find in later steps of FA. You will even find some of them within departments as you work on the subfunctions of each unit. Happy hunting!

13 Information Flowchart

INTRODUCTION

The information flowchart is a good way to show the logical relationships of subfunctions. It places specific emphasis on the flow of information, but it is also able to show the flow of material and services among business units. It shows the flow both within a department and between departments.

The information flowchart is a good vehicle for aiding the Functional Analysis (FA) team to locate discontinuity. That is, it helps to find products without a customer and customers without a source of the products they need to perform their functions. It requires a lot of teamwork and the exchange of ideas among the team members if the information flowchart is to serve its purpose. Where the team finds discontinuity, they have also found a conflict that must be resolved.

DICTIONARY DEFINITION

The word *information* is defined by *Webster* as "news or intelligence communicated by word or in writing." *Webster* defines the word *flowchart* in a way that is very close to the needs of FA. It is "a detailed or graphic representation, using symbols, which illustrates the nature of and sequencing of an operation on a step by step basis."

WORKING DEFINITION

In FA an *information flowchart* is a chart that shows the flow of information, material, and services among the various subfunctions of a business. The emphasis of the flowchart is placed on information. But related materials and services are also charted, since the information often is for the purpose of planning or ordering these other products.

IDEFO

IDEFO is a flowcharting technique well suited to the purpose of FA. It was developed by the Department of Defense (DOD) as a vehicle for documenting how a portion of a business operates. It was not meant to deal with information specifically but with all aspects of the business. IDEFO is a more detailed and rigorous technique than is needed for the use of FA. The rules for mapping information flow in IDEFO are quite precise. But the method of obtaining the knowledge about the information flow to be mapped is not supported by an analytical technique of corresponding rigor.

In a simplified form, IDEFO is quite useful for FA. It gives FA a simple visual model of the flow of information within the business. FA supplements the IDEFO style of flowchart with a text that describes when and how the flow takes place. The combination is very effective in identifying and resolving conflict among business units.

There are a number of commercial programs for personal computers that you can use to prepare the flowcharts. Their greatest value is in making it easy to change the charts as you learn new things.

FLOWCHART CONVENTIONS

Refer to Figure 13–1 for an example of the flowchart conventions for a single subfunction of a department. The name of the subfunction is printed in the box that represents the subfunction.The output flow line exits the right side of the box and terminates at

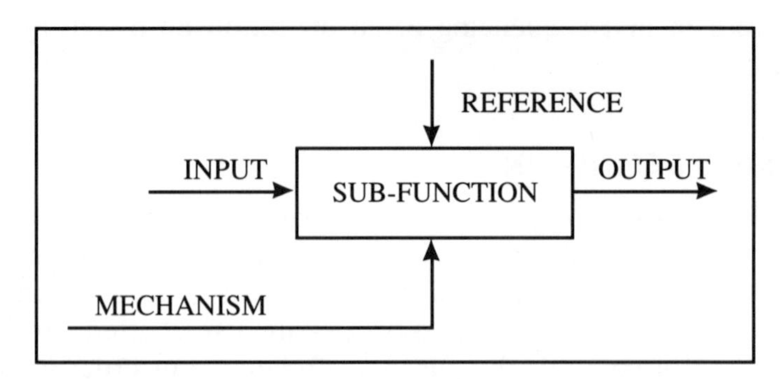

Figure 13–1. Information flowchart conventions.

some other box. The flow line represents a product of the subfunction. A single subfunction will have at least one output but can have more than one. The input flow line enters at the left side of the box and represents something that is used by the subfunction and is transformed into the output. Input can be such things as raw material or data. There need not always be an input, but there usually is at least one.

The task performed by the subfunction might need reference materials such as standards or schedules. These are represented by a flow line that enters at the top of the box. These references might or might not exist for any specific subfunction. The subfunction sometimes needs one or more mechanisms in order to perform its task. This might be people or tools, or other resources. These are represented by a flow line that enters at the bottom of the box.

It is usually best to have a single flowchart per department when doing this step of the FA. This keeps the volume of detail on a single page to a manageable size of three to six subfunctions.

EXAMPLES

The flowcharts shown in Figures 13–2 through 13–7 show the same example departments that were discussed in previous chapters of this text. Note that in the examples some of the outputs of subfunctions are not inputs to other subfunctions but take other forms such as reference material.

In Figure 13–2, the manufacturing schedule, the product of the first box, is reference material for the other two subfunctions, not input. It is not changed by these other subfunctions but is used by them as a guide to their activity. In Figure 13–2, the third box has two inputs, raw production materials from some unidentified source and components that are output or products of the subfunction called PRODUCE COMPONENTS.

In Figure 13–3, the subfunction called STORE MATERIAL produces a bin catalog but also uses it as input, since it maintains a record of the location of the material that it stores.

In Figure 13–4, the subfunction called ESTABLISH CONTRACT needs no input, since nothing is physically changed when it sets up a supplier contract. Of course, a piece of paper is changed into a contract, but this is too trivial to document with an input called "PAPER."

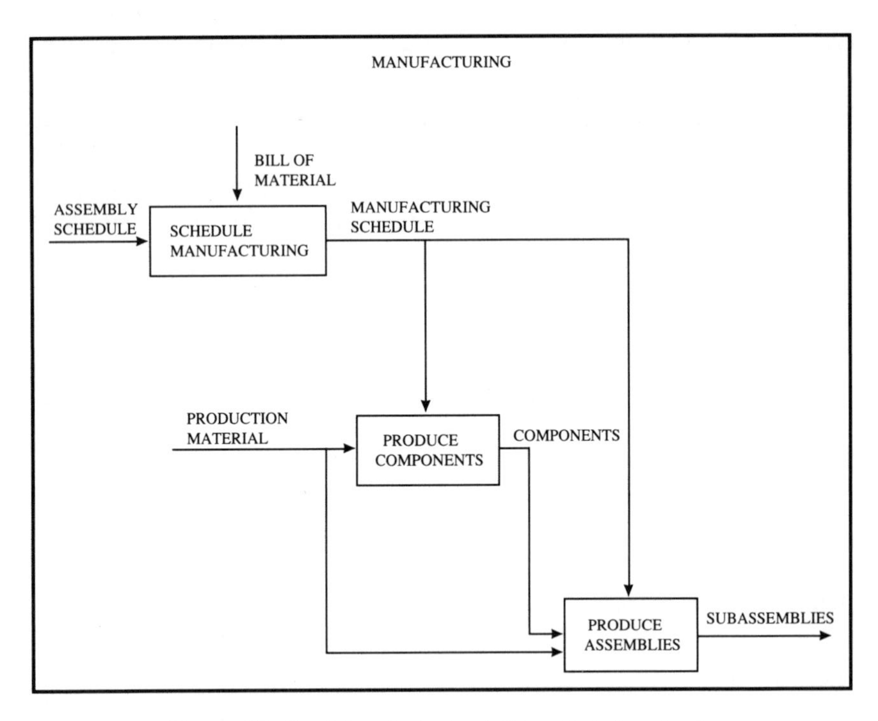

Figure 13–2. Manufacturing information flowchart.

In Figure 13–7, the assemblers serve as a mechanism, since they are needed to perform the task of the subfunction. Mechanisms are not normally shown in a flowchart unless they must specifically be planned, ordered, and provided by some other subfunction.

METHOD OF REPRESENTATION

Figures 13–2 through 13–7 show the way in which the flowcharts should be drawn. There is a chart for each department and a box on the chart for each subfunction within the department. The boxes are staggered from the upper-left corner of the page to the lower-right-hand corner.

It is most convenient to sequence the boxes in a logical order, as in the sequential flow of information and material within the department. This is how it is done in Figure 13–2. This flow toward the lower right makes it easier to draw the flow lines and reduces the need for crossed lines to represent flow back to a box toward the upper left.

Notice that the flow of material and information between depart-

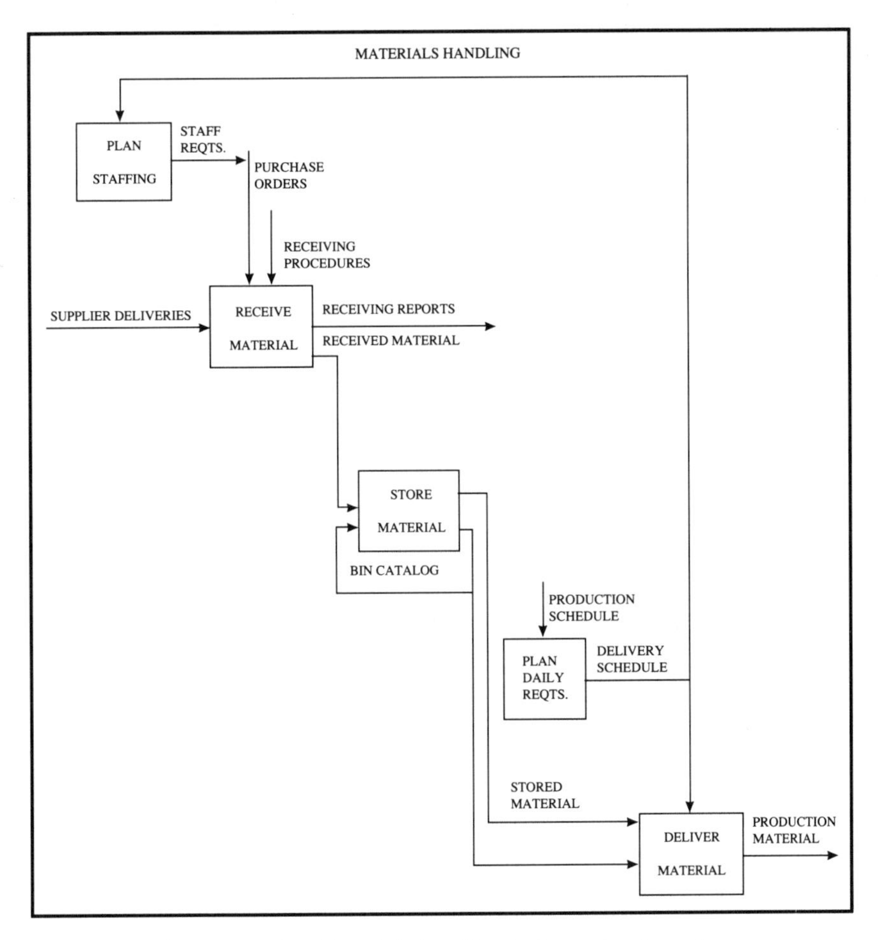

Figure 13–3. Materials handling information flowchart.

ments is not shown in completed form on the flowcharts. It is harder to identify the end points of the flow outside the individual departments. Where is the production schedule on Figure 13–5 used? In how many places is it used? As the analysis of subfunctions goes into more detail in the next few chapters, the source and sink (end point) of each flow will be identified.

ANALYTICAL METHOD

The steps for developing an information flowchart are shown in Figure 13–8. The building block of a departmental chart is the subfunction. You have prepared the basis for this in Chapter 10.

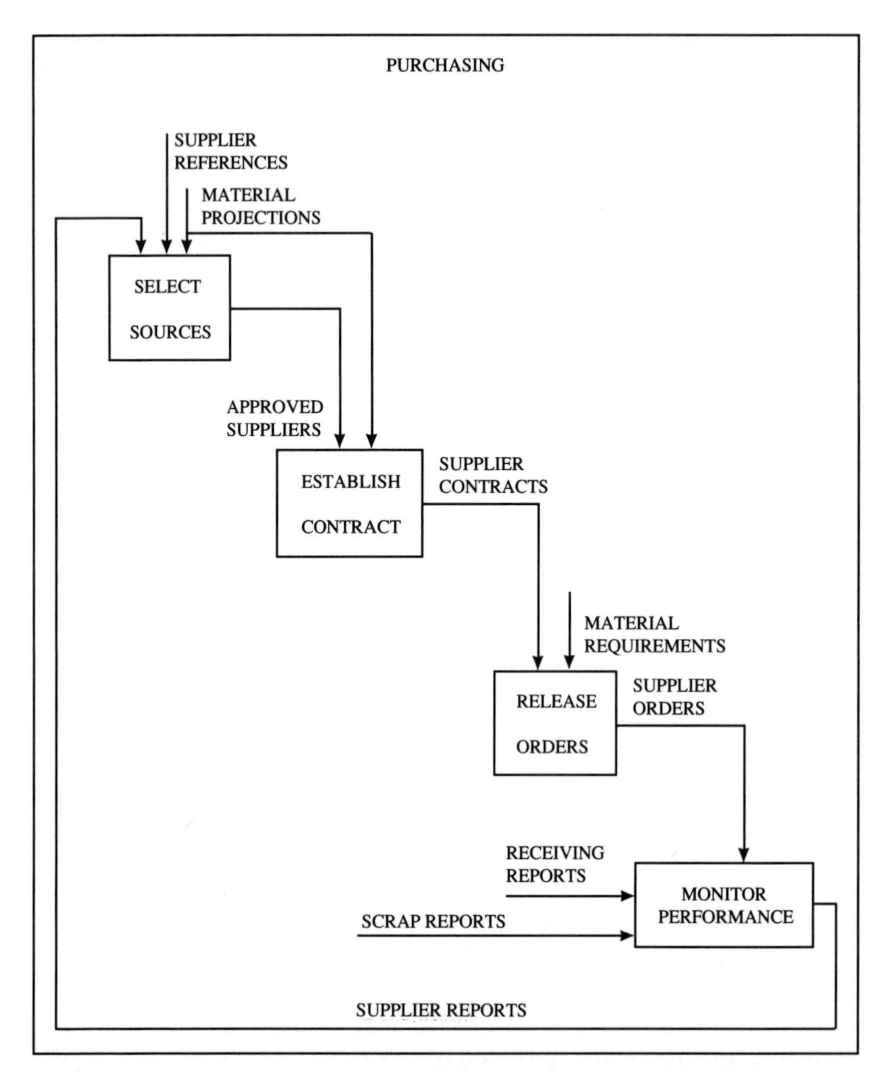

Figure 13–4. Purchasing information flowchart.

You might, however, have found the need to change some of the subfunctions in later steps. If you have less than three or more than six subfunctions for a given department, give serious consideration to redefining the departments. Remember that you are doing an analysis and are not reorganizing the company.

Try to determine the logical sequence of information or material flow from one subfunction to another. The main product of each

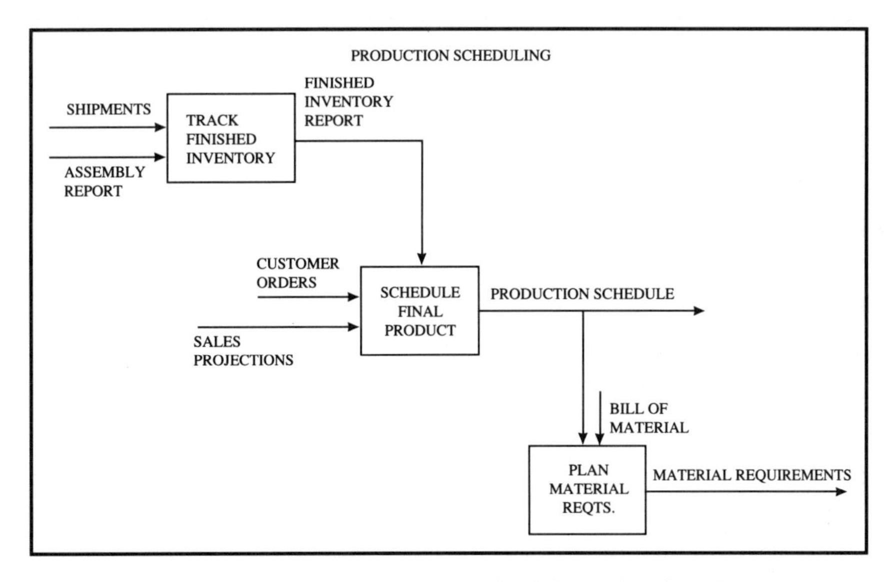

Figure 13–5. Production scheduling information flowchart.

subfunction should be obvious and will usually be time dependent from subfunction to subfunction.

Ask what is produced in each subfunction. What is the main product? What are the related products? If the main product is material or a service, there is often an associated set of information, such as a production report or a scrap report.

Ask what is needed in order to produce the output. Most of the time there is some sort of input or raw material from which the output is formed.

There is often some sort of order or schedule that is the thing that prompts the subfunction to perform its task. There is often a recipe or a work order or a set of standards to guide the process. On occasion, there is a significant mechanism.

Do not worry too much about whether the output is used by another department or by an end customer. Consideration of that will come later in the study. Each department will find it easier to identify its own needs than to identify the needs of other units for the products that it supplies.

Since there is a source and a sink for each product, it only requires one of the two to identify the need for the product. If a subfunction does not know where a necessary input, reference, or

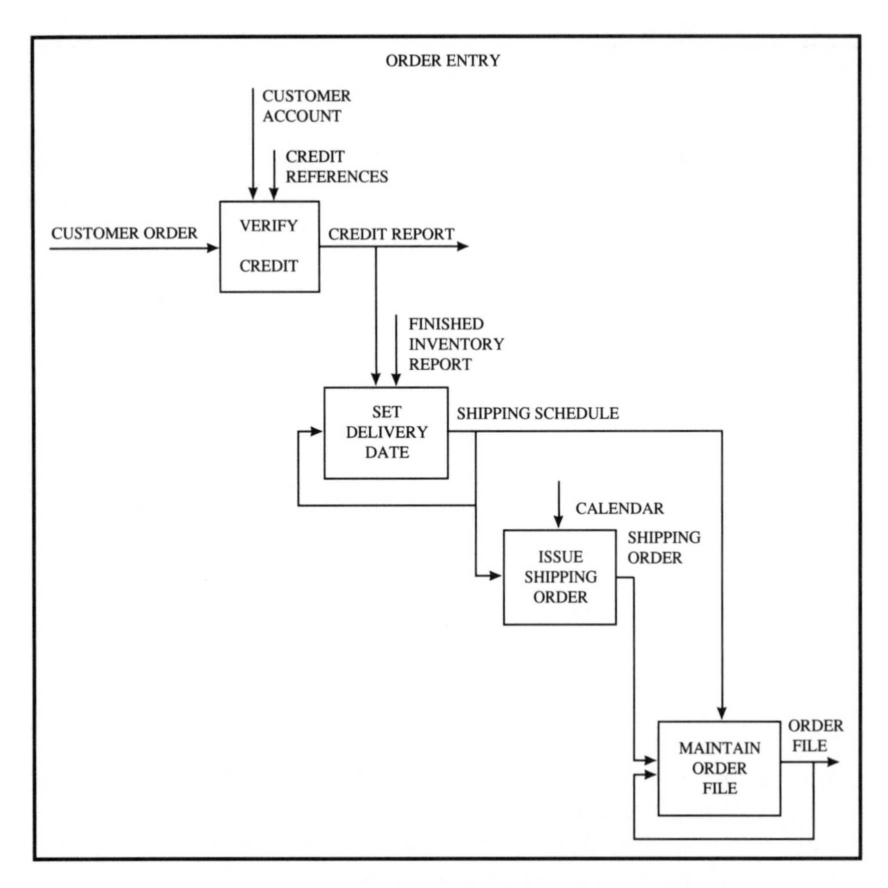

Figure 13–6. Order entry information flowchart.

mechanism comes from, talk to other departments that might be sources. If it is needed, someone must be in a position to produce it. But do not work too hard at finding all the sources and all the users of each output. That will be done in the next chapter.

A department should at this time identify anything that it needs but does not receive. This is especially the case if the department is changing a basic strategy. For example, if there is a move to JIT production, then there is most likely the need to guide delivery of materials by a sequential schedule of production.

Feel free to reorganize the subfunction breakdown if the original one becomes inconvenient. You can make any changes you wish in

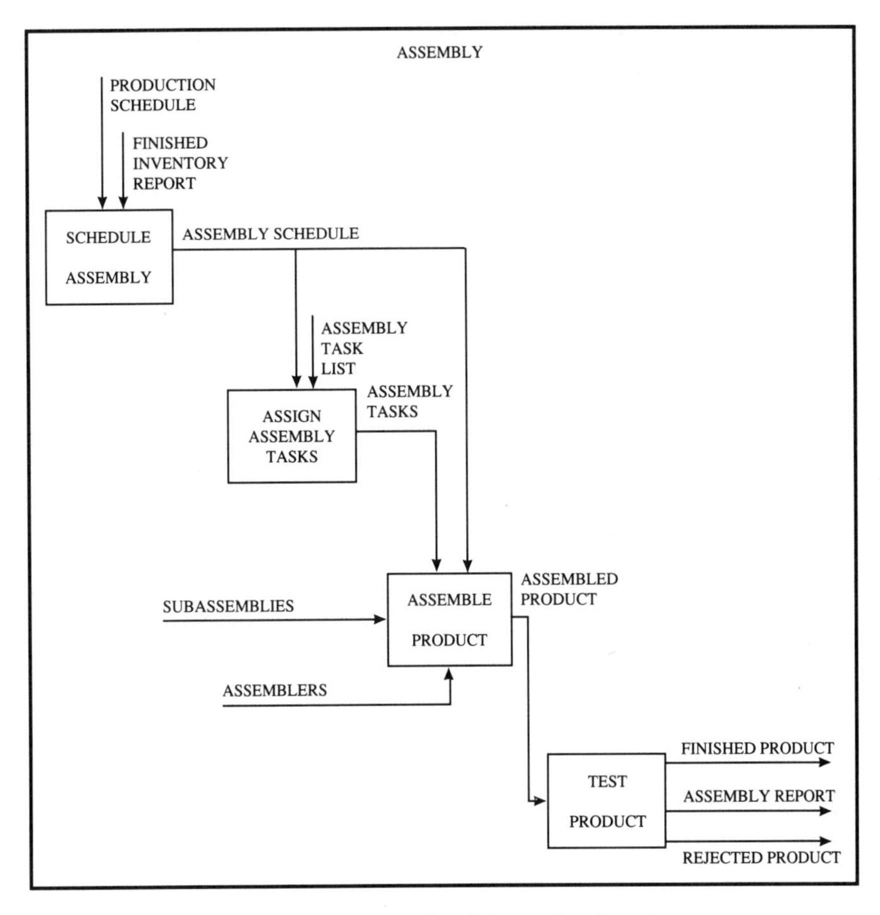

Figure 13–7. Assembly information flowchart.

the form of additions, deletions, or even more detailed breakdown of the charts. Change any labels you wish in order to make them as meaningful as you can. The label on a box should in some way tell that the purpose of the process within the box is to produce the main output of the subfunction.

The subfunctions need not represent units of organization but the responsibilities that are performed within the organization. An individual person or a group of people could perform more than one of the subfunctions. It all depends on how you wish to represent things.

1. Start with subfunctions
2. Determine sequential flow
3. Ask what is produced
4. Ask what is needed to produce output
5. Ask when it is produced
6. Iterate as necessary
7. Record conflicts

Figure 13–8. Information flowchart analytical method.

MORE CONFLICTS

As you draw the charts, you will most likely find a few more conflicts within your department or with other departments, just as you found them in the step described in Chapter 12. Document these conflicts in the same manner as you did then.

The set of documented conflicts grows at each stage of FA until they are resolved, usually by changes in operational rules. This takes place in Chapter 15. The parallel documentation of conflicts is shown in Figure 2–3. There is no single best time to identify conflicts. They come up repeatedly as you perform the steps in Chapters 11 through 14.

Figure 13–9 shows an example of a conflict found during the development of the charts. How does Materials Handling know what material to deliver? How does it know what quantity is needed? There is no mechanism for Assembly or Manufacturing to use in order to tell Materials Handling that there is an exception to the schedule.

One more conflict found while developing the charts is shown in Figure 13–10. Materials Handling has a subfunction to plan staffing, but there is no indication that the information is used by anyone in order to provide the staff.

It is apparent that the term *conflict* should be used in the broadest

CONFLICT

Between: Materials Handling
Manufacturing and Assembly

Subject: Delivery of short production materials

Cause: Lack of a defined mechanism for Materials Handling to be aware of a material shortage

Solution: Short reports will be made by Assembly and Manufacturing upon identification of any scrap or loss of material

Delivery will be made within 30 minutes

Figure 13–9. Resolved information flow void.

CONFLICT

Between: Materials Handling
Personnel

Subject: Staffing requirements

Cause: No communication has yet been made between the two departments to determine how staff will be provided to accommodate workload fluctuations

Solution: Optional solutions to be developed

Henry Hardy of Materials Handling
Herman Hyre of Personnel

Due April 15, 1991

Figure 13–10. Unresolved communication void.

sense. This includes "unanswered questions" as to the source that must provide the needed inputs, references, or mechanisms. It also includes the analysis of those outputs that nobody seems to need.

MATERIAL CARRIES INFORMATION

As stated early in the book, there is a tendency to use computers more than is needed. One result of this is that information is often thought of as existing only in mechanical or printed form.

A little understood fact is that each piece of material can carry information with it. Returned scrap material picked up by Materials Handling can be the signal for the delivery of the same quantity of fresh material to the subfunction that made the scrap. The Japanese sometimes place a sequencer between a manufacturing cell and an assembly line. The device is no more than a rack to hold material for assembly use. The unit that makes the material can tell what finished parts are needed. It does this by the fact that labeled spaces in the rack are empty of specific parts. The information "bring more parts" is inherent in the empty space. The label of the parts that are needed is on a sign attached near the empty space. That is where the parts are always to be placed.

COMMON ERRORS

A common error is to try to place too much detail on an information flowchart. This results in a crowded and confused chart that carries a lot of data but little intelligible information. A reverse sort of error is to provide too little detail, which would be the case if there were no new information represented on the chart. This results from trying to avoid the work needed to think through the chart, rushing the job.

It is an error to think that everything is either input or output. Many of the things most needed to perform a function are reference items. And occasionally, there is a need to identify a mechanism. An example might be seen in a subfunction that we can label "PLAN PROCESSES." If the reference material is in the form of printed standards, the flow line comes into the top of the box. But if the process is done by mechanical means, such as through a computer program that automatically generates the plan, then the chart might not show any reference material but a mechanism, a computer program, shown by a line entering the bottom of the box.

Despite the use of a mechanism in the above example, it is an error to pay too much attention to mechanisms, which are usually not vital to the analysis of information flow. If you chart too many mechanisms, you will fall into the error of attempting to put in too much detail.

The last serious error is to fail to discuss the flow of data among the team members. This will result in disjointed charts that do not accurately portray the flow of the business.

CAUSES AND CURES

One major cause of excessive detail is "nitpicking." The analysis task can never be delegated to a person who is overly concerned with detail. This too often is the person who is volunteered to be a member of the team. After all, he or she is gladly spared from other work. There is a need to focus on the essence of the matter. This requires first-class minds that can distinguish the vital from the trivial.

As before, a common cause of poor analysis is cursory treatment. The analysis needs to be assigned to a person who honestly cares to do a good job and believes in the value to be obtained from the effort.

Lack of teamwork is another common cause of poor results. Although team meetings are not necessary at this stage, the effort spent in communication and discussion among the team members helps develop valid and usable charts.

Another cause of poor flowcharts is a failure to ask "when." Put yourself in the place of a person performing a subfunction. For each product, you need to ask yourself if you know how and when the provider of the product finds out that you need it. What is the signal that tells him or her that you are ready to receive the product and, indeed, when you need it in order to perform your function? Now, as long as you have put yourself in the place of the person doing the function, also ask yourself how you know when your customer needs your product. And ask how you might know how many the customer needs—and when they are needed.

QUALITY CHECKLIST

Figure 13–11 shows a checklist for you to use to verify the quality of your information flowcharts. There should be three to six boxes

1. Three to six subfunction boxes
2. Less than four arrows on any side of a box
3. Less than eight arrows with any box
4. Minimal crossing of arrows
5. All arrows labeled

Figure 13–11. Information flowchart quality checklist.

representing subfunctions. Less than three makes the chart trivial. It would be better to include it in another "department." More than six makes it too busy to be easily understood and makes the flow of lines confusing. Break it up into two or more departments.

There should be no more than four arrows or flow lines on any side of a subfunction box. If there are more, you have made the chart confusing. Either you have fallen into the error of putting in too much detail, or you have tried to merge several subfunctions in one box. There should be no more than eight arrows in total on a box. The reasons for this are the same as above: The chart should be simple to read.

There should be little crossing of arrows. The purpose of this is to avoid confusion in following the flow of information. If there are a lot of crossed lines, you should look to see if you might have placed the boxes out of time-frame sequence. You might be able to interchange several of the boxes to make the flow simpler to follow.

All arrows should be labeled. Each flow line represents a product that has a source and a use. The product is important, so it should be titled.

SUMMARY

The information flowchart is a handy, visual device for showing the flow of information, material, and service within a business. It is a vehicle for aiding in the analysis of how information is triggered, prepared, and transmitted.

The volume of documentation that you have produced in the FA has grown and will now grow significantly larger. For this reason, it is necessary to have a summary-level visual reference to keep the information organized. The flowchart serves this purpose.

The charts begin to force more communication among organizations. Since the flow lines have two ends, they must begin in a subfunction and end in one or more other subfunctions. The flow line that is left hanging is a symptom of incomplete analysis. And it is possibly a sign of poor operating procedures within the business, a product without a source or without a user. This will be resolved in the next chapter, the next stage of FA.

14

Subfunction Description

INTRODUCTION

At this stage of Functional Analysis (FA) you begin the first really detailed work of information analysis. The need for regular discussion among team members from the various functional areas is vital to progress from this point on. This step should lead to the discovery of a lot of conflict over the flow of product among organizational units. It should also begin to give you hints of the changes that might be made to reduce the conflict or to eliminate it.

DICTIONARY DEFINITION

Webster defines the word *description* in this way: "A description is an account of the properties or appearance of a thing, given so that another person may form a concept of it."

WORKING DEFINITION

In FA, a subfunction description is a short, single-page summary of key information that gives a perspective on the method of operation of a subfunction. It tells the most important facts about what the subfunction does, how it does what it does, and what makes it do it.

ITEMS IN A DESCRIPTION

The key items in the subfunction description are those factors that impact on the product that the subfunction makes and how it is made. They are shown in the list in Figure 14–1 and are described here by title:

- *Function* is really a mission statement that begins with a verb and identifies the main product of the subfunction.
- *Trigger* is that which causes the subfunction to do its assigned

1. Function
2. Trigger
3. Response
4. Horizon
5. Outputs
6. Inputs
7. References
8. Mechanisms

Figure 14–1. Subfunction description items.

task. It could be the arrival of information such as an order, the arrival of material to be processed, or the arrival of a calendar date such as a month end. The task is performed in response to an expression of need by the user.

- *Response* is how quickly the subfunction reacts to the trigger. It is the normal time frame that is needed to provide the needed product for the user.
- *Horizon* is how far into the future the subfunction looks when providing material or information needs. This could be the number of weeks or months that are planned for in a production schedule. It could be how far in advance a supplier is notified that raw material is needed.
- *Outputs* are from the flowcharts in Chapter 13, with a short label to show the box on the flowchart where each output is used.
- *Inputs, References, and Mechanisms* are dealt with in the same way as the outputs, with a short label on the flowchart to show where each comes from.

EXAMPLES

A number of examples of subfunction descriptions are shown in Figures 14–2 through 14–6. Each example represents the detail

	PLAN STAFFING	MH1
Function:	Determine the quantity of labor required to receive and deliver production materials for the near future	
Trigger:	Receipt of Production Schedule	
Response:	Staff requirements calculated within one working day	
Horizon:	Staffing requirements are projected for the horizon of the Production Schedule - the following calendar month	
Outputs:	Pe3 - Staff requirements	
Inputs:	None	
References:	MH4 - Delivery Schedule	
Mechanisms:	None	

Figure 14–2. MH1—Plan staffing.

behind one of the subfunctions of Materials Handling that is charted in Figure 13–3.

METHOD OF REPRESENTATION

There is some brief coding that is needed for the subfunction descriptions. Note that the pages have been given an identification label in the upper-right-hand corner. The MH1 in Figure 14–2 refers to the first subfunction of Materials Handling as diagrammed in Figure 13–3. The same MH label has been added to the upper-right-hand corner of the original flowchart. See Figure 14–7. The upper-left- hand box on the chart is MH1, the next is MH2, and so on.

The flowchart has also been updated with the addition of the proper code for each flow line, shown in Figure 14–7. If the flow line terminates on the same chart, it has no such label. This is the case for stored material that moves from MH3 to MH5. But if the flow line connects with a box on another page, it is given a label to show

RECEIVE MATERIAL	MH2
Function:	Process material deliveries from suppliers
Trigger:	Arrival of supplier delivery vehicle
Response:	Material removed from receiving area within one hour of arrival
Horizon:	Not applicable
Outputs:	MH3 - Received material *** - Receiving Reports
Inputs:	INP - Supplier deliveries
References:	Pu3 - Supplier Orders QC3 - Receiving Procedures
Mechanisms:	None

Figure 14–3. MH2—Receive material.

where it terminates. Receiving reports from MH2 move to QC2, the second box on the chart for Quality Control (not included in examples). The products known as production material are the output of box MH5 and move to three other boxes in Assembly (As2) and Manufacturing (Ma2, Ma3). The input, supplier deliveries, to MH2 is labeled INP since it is input from outside the business.

Note the word "None" after the title "Mechanisms:" in Figure 14–2. This is a term used when there is nothing to list. The label "Not applicable" is also used after the title "Horizon:" in Figure 14–3 when there is no time horizon for the function.

ANALYTICAL METHOD

The steps in the analytical process are listed in Figure 14–8. Label each department on the flowchart, in a manner similar to the "MH" for Materials Handling. The label can be whatever you wish. A mnemonic label seems best as a memory aid.

	STORE MATERIAL MH3
Function:	Provide temporary warehousing for production material until it is needed in the production process
Trigger:	Received material deposited at the warehouse
Response:	Material stored within 15 minutes of completion of the receiving process
Horizon:	Not applicable
Outputs:	MH5 - Stored Material MH3 - Bin Catalog
Inputs:	MH2 - Received Material MH3 - Bin Catalog
References:	None
Mechanisms:	None

Figure 14–4. MH3—Store material.

Trace the flow of input and output, references, and mechanisms to find out where they begin and end. For each flow line that leaves a page, label it with the proper code to show the corresponding subfunction in another department. Or label it with INP or OUT to show it as something that enters from outside the business or that passes outside the business, as in an order to a supplier.

A mnemonic label is not shown on the flowchart for a product that does not flow to or from another department. But the proper label should be shown on the subfunction description. Figure 14–3 shows received material as an output to MH3. The label "MH3" helps the person who is reading the subfunction description. But the label "MH3" is omitted from the chart on Figure 14–7 since the location of box MH3 is obvious.

PLAN DAILY REQUIREMENTS	MH4
Function:	Determine the schedule for regular delivery of production materials to meet the Production Schedule
Trigger:	Receipt of Production Schedule
Response:	Delivery Schedule calculated within one working day
Horizon:	The Delivery Schedule is calculated for the horizon of the Production Schedule - the following calendar month
Outputs:	MH1 - Delivery Schedule MH5 - Delivery Schedule
Inputs:	None
References:	PS2 - Production Schedule
Mechanisms:	None

Figure 14–5. MH4—Plan daily requirements.

You will need to resolve discrepancies with neighboring departments where an output is not used or where an input, reference, or mechanism has no origin. Every flow line must begin and end somewhere. That is one nice thing about the flowchart: It keeps you honest. If you make sure that the flow lines have no loose ends, you will identify voids and conflicts. Note that the output "RECEIVING REPORTS" of subfunction MH2 has no label in Figures 14–3 and 14–7. The using subfunction has not yet been identified.

Change any inconsistent labels. An example is the product called "purchase orders" in Figure 13–3 but called "supplier orders" in Figure 13–4. Many times the labels on products will be inconsistent since the charts and the subfunction descriptions have been developed separately by the different units. If the chart is to be your road map, the roads must be clearly labeled.

	DELIVER MATERIAL MH5
Function:	Move production material to the site of the production processes as required
Trigger:	Delivery Schedule
Response:	All deliveries completed within 15 minutes of the scheduled time
Horizon:	15 minutes, maximum
Outputs:	As2 - Production Material Ma2 - Production Material Ma3 - Production Material
Inputs:	MH3 - Stored Material
References:	MH4 - Delivery Schedule
Mechanisms:	None

Figure 14–6. MH5—Deliver material.

You will need to develop text to describe in brief but clear terms what is meant by *function, trigger, response,* and *horizon*. What function does the unit perform? What makes it act? How does the subfunction know when to act? How does it know how much product to make? How long does it take to act? And how far into the future does it look when it acts?

Find out how a subfunction conveys its needs to other subfunctions on which it depends. Ask the representative of the supplying department if he or she agrees on the format and timing of the signal of need for the product. Ask what the using subfunction expects in terms of product quality and timeliness. Does the user pay for the products it uses? If so, what price is he or she willing to pay? Ask what happens when something goes wrong, as when faulty parts are delivered by a supplier. Do the parts still get stored in the warehouse, or are they returned to the supplier? How is the

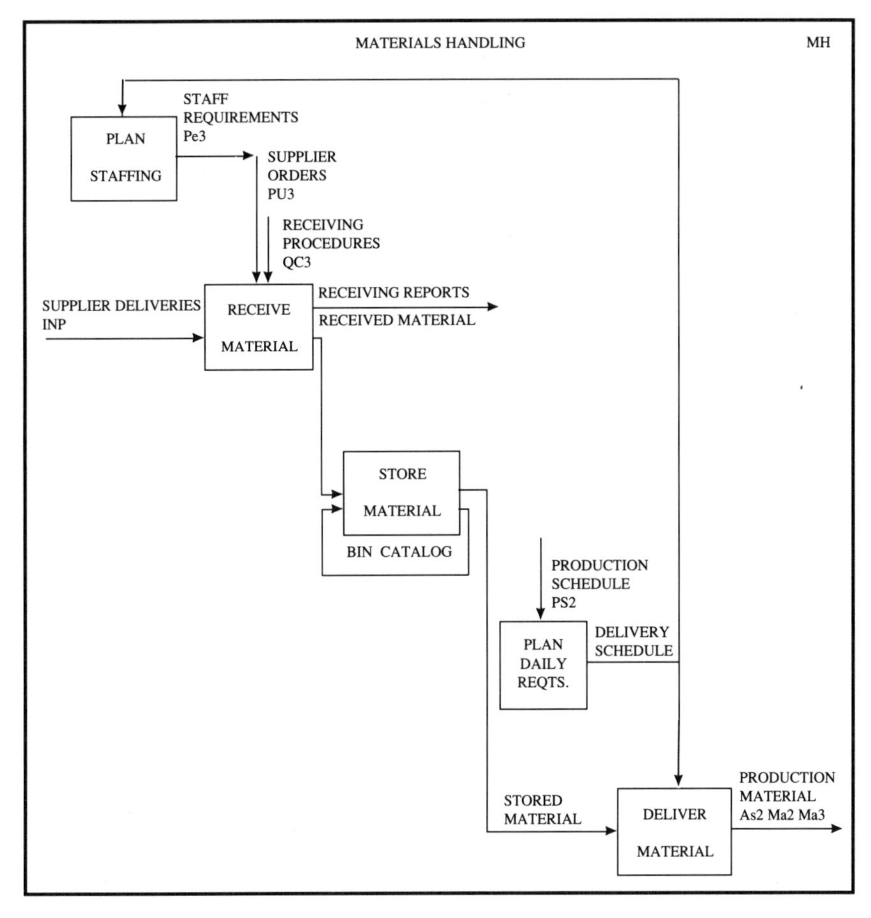

Figure 14–7. Materials handling information flowchart.

shortage in material made up? Is another order placed with the supplier?

Systems would be very simple if nothing ever went wrong.

COMMON ERRORS

It is an error to assume that you already know what other departments need from your department or to assume that they already know what yours needs from them. It is best to compare notes with them as you develop your information flowcharts.

It is an error to try to include too much detail in the charts. If the

1. Label each department
2. Trace flow
3. Label each off-page flow
4. Resolve discrepancies between subfunctions
5. Change inconsistent labels
6. Develop descriptions of
   - function
   - trigger
   - response
   - horizon
7. Ask how each subfunction is triggered
8. Ask how each sends a message of need
9. Ask quality, timeliness, cost expectations
10. What happens when something goes wrong?

Figure 14–8. Subfunction description analytical method.

charts become crowded, they will not be used. The charts are intended to aid communication, not to impede it. It is equally an error to go into too little detail. If the charts do not show a comprehensive picture of the flow of information, they will not help you to find and resolve conflict.

Do not be casual about triggers. They are very important and are often overlooked when planning for information systems. A common problem with CIM and JIT is the failure of planners to take into account how the needs of the user are to be made known to the supplier.

You can also be too casual about horizons. Many of the conflicts between units come from an incompatibility of planning and delivery horizons, which is illustrated in the conflict shown in Figure 12–4.

CAUSES AND CURES

The most common cause of error was illustrated above. It is the failure to talk to your neighbors about what they need from you and what you need from them. You must have this exchange of needs description sooner or later. Best to do it now. The exchange of ideas and needs is, of course, more likely to take place if there is top management support of the FA project. An occasional pep talk from the sponsor can be a good thing.

As usual in FA, things tend to slow down if there is a lot of emphasis on excessive detail. And again, this can be avoided if you can keep from having detail-minded people on the team. The "brush-off" is always wrong. Do not give the analysis task to the person who can be spared. Give it to the person who has the skill and the interest. The people on the team must want to do the job. They must be motivated.

You can get bogged down in conflict if you are not careful. The same advice applies here as it did in Chapter 12: Do not try too hard to solve problems yet—just document them and assign responsibility and a target date for when the solutions need to be in hand. Of course, if a solution to a conflict is obvious, then you should document it, too.

QUALITY CHECKLIST

Refer to Figure 14–9 for a checklist by which to test the quality of your flowcharts. Remember that you are preparing two different types of document: The chart is an updated copy of the information flowchart; the other document is a subfunction description for each box on each flowchart.

So the first thing to verify is that you have a description page for each subfunction box on the flowchart. Make sure that each flow line that goes off the page of a flowchart is tagged with the department and box of the subfunction to which it refers. Also make sure that the product that is represented by the flow line is shown as an

1. A description page for each box
2. A label on each off-page flow
3. Triggers clearly identified
4. Horizons clearly identified
5. Exceptions/conflicts documented

Figure 14–9. Subfunction description quality checklist.

input, output, or other product on the subfunction description. Are its title and label the same on all documents?

Be sure that the triggers are described clearly and that they make sense to both the owner of the subfunction and the one who sends the trigger. Are the horizons identified and described so that they are readily understandable? Both the horizons and the triggers are often overlooked in superficial approaches to information analysis.

Are all exceptions identified and documented? If you have completed this phase of FA, all exceptions should now be documented as conflicts.

STILL MORE CONFLICTS

This phase of FA should bring to light a great number of new conflicts that were not seen before. Figure 14–3 is a case in point. What should Materials Handling do with received material if Manufacturing or Assembly is in short supply? Should it be stored in the warehouse as usual, or should it be forwarded to the point of need? Does Materials Handling know that it is needed? How does it know? One more possible conflict is shown in Figure 14–3: Who uses the receiving reports? They do not seem to have a user. Does Accounting need them in order to verify receipt before it pays the bills?

As you look at Figure 14–7, you might begin to wonder how the production schedule is developed. To learn this you would have to look at the information flowchart for Production Scheduling. You might also look at the detail in the PS2 subfunction description. One reason why you might want to do this is to find out how scrap is handled. Is it reported? Is it taken into account when setting the production schedule? Figure 14–7 shows production material being delivered to As2, the assembly of product. If you refer to Figure 13–7, you will see that the need for these materials was overlooked on the original chart. This oversight was easily corrected at this stage of FA. It was not really a conflict but would have become one if not corrected.

SUMMARY

This is the first step of FA that has required a lot of communication and cooperation among business units. It is the prelude to the next step, which is the definition of rules of operational behavior. These are the means by which the conflicts will be resolved. Some of the rules will be straightforward. They are merely a listing of the common practices of each subfunction. These rules complete the description of the subfunction. But some of the rules will be in response to identified conflicts. These are harder to define. They will take even more teamwork among the various departments, as well as a lot of imagination.

15 Rules to Resolve Conflicts

INTRODUCTION

At this stage of Functional Analysis (FA), all conflicts must be reconciled. In previous chapters, you have been told that you could put off resolving the conflicts. The buck stops here. This is the tough stage in which the team members must work together as never before.

The rules of operation of each subfunction constitute the vehicle for resolving the conflicts. But sometimes the change in the rules is quite significant in that it denies a previously stated strategy or denies a measure of performance. Thus, it is possible that there will be a need for iteration to change some of the strategies or measures—or anything else in between.

This stage requires teamwork and more teamwork. Remember that your task is to improve the operation of the business as a whole.

DICTIONARY DEFINITION

As usual, *Webster* has a definition to offer: A *rule* is "an established principle, a standard or guide for action."

WORKING DEFINITION

Rules, in the context of FA, are the principles of action of a subfunction that ensure that the products of the subfunction serve the needs of the users. Rules are also the vehicle for resolving organizational conflict.

RULE TOPICS

The topics to be considered in rules are shown in Figure 15–1. They are:

1. Who does it
2. When it is done
3. How quickly it is done
4. Usability of the output
5. Cost of the product

Figure 15–1. Rule topics.

- Who is the one who is responsible to perform a specific function? This is usually implicit in the label of the department and of the subfunction.
- When is the function to be performed? This is usually shown in the trigger statement.
- How quickly does the task need to be done? How timely is the output available to the user? This is a key measure of the product and is usually shown in the response statement.
- What is the quality of the product? How usable is it? This is usually shown in the rules that tell how the product is prepared.
- What is the cost to make the product? This is shown in the rules.

EXAMPLES

The examples in Figures 14–2 through 14–6 have been completed by the addition of appropriate rules. These are shown here as Figures 15–2 through 15–6. The number of rules is small since each subfunction has a limited number of products and since the response and horizon topics contain implied rules. Rule 1 in Figure 15–2 helps to ensure a quality of product adequate for the Personnel Department to plan the acquisition of staff. Rules 3 and 5 in Figure 15–4 are for the purpose of maintaining control over the quality of the product. Rule 3 in Figure 15–5 is a cost control measure, a way to balance quality and cost.

PLAN STAFFING MH1

Function:	Determine the quantity of labor required to receive and deliver production materials for the near future
Trigger:	Receipt of Production Schedule
Response:	Staff requirements calculated within one working day
Horizon:	Staffing requirements are projected for the horizon of the Production Schedule - the following calendar month
Outputs:	Pe3 - Staff requirements
Inputs:	None
References:	MH4 - Delivery Schedule
Mechanisms:	None
Rules:	1. The staff requirements distinguish full-time employees from part-time
	2. The staff requirements are separated by labor grade
	3. The staff requirements for part-time are smoothed by week

Figure 15–2. MH1—plan staffing.

METHOD OF REPRESENTATION

Figures 15–2 through 15–6 are good examples of how rules add to the description of subfunctions. They serve to provide a fuller documentation of how the function is to be done. They show the means by which the product is to be made so that it will serve the quality and timeliness needs of the user.

RECEIVE MATERIAL	MH2
Function:	Process material deliveries from suppliers
Trigger:	Arrival of supplier delivery vehicle
Response:	Material removed from receiving area within one hour of arrival
Horizon:	Not applicable
Outputs:	MH3 - Received material Ac3 - Receiving Reports
Inputs:	INP - Supplier deliveries
References:	Pu3 - Supplier Orders QC3 - Receiving Procedures
Mechanisms:	None
Rules:	1. Material is checked for count and quality based upon Receiving Procedures 2. Material flagged as urgently needed is set aside for special delivery 3. Count errors are noted on the copy of packing list 4. Below-standard materials are set aside for Quality Control with a note of the problem placed on the packing list

Figure 15–3. MH2—receive material.

It is hard to tell you any more clearly what should be included in rules and what should be excluded. You can get carried away with many rules that really have no importance. The thing you should ask is whether or not the rules omit anything important that would prevent the product from meeting the user's expectations.

STORE MATERIAL MH3

Function:	Provide temporary warehousing for production material until it is needed in the production process
Trigger:	Received material deposited at the warehouse
Response:	Material stored within 15 minutes of completion of the receiving process
Horizon:	Not applicable
Outputs:	MH5 - Stored Material MH3 - Bin Catalog
Inputs:	MH2 - Received Material MH3 - Bin Catalog
References:	None
Mechanisms:	None

Rules:

1. Like materials are stored close together where possible
2. Old materials are consolidated when workload is light
3. Audits of storage contents and counts are performed on the basis of item value and age in storage - when workload is light
4. The bin catalog is posted after each unit of product is handled
5. Oldest material is picked for a delivery, and partial containers are picked before full containers

Figure 15–4. MH3—Store material.

PLAN DAILY REQUIREMENTS	MH4
Function:	Determine the schedule for regular delivery of production materials to meet the Production Schedule
Trigger:	Receipt of Production Schedule
Response:	Delivery Schedule calculated within one working day
Horizon:	The Delivery Schedule is calculated for the horizon of the Production Schedule - the following calendar month
Outputs:	MH5 - Delivery Schedule *** - Delivery Schedule
Inputs:	None
References:	PS2 - Production Schedule
Mechanisms:	None
Rules:	1. The delivery quantity is to be within the physical storage capacity of the using area
	2. The Delivery Schedule is to smooth the delivery staff requirements as much as possible
	3. 10 percent monthly staffing allowance is to be made for load consolidation and record verification

Figure 15–5. MH4—plan daily requirements.

DELIVER MATERIAL	MH5
Function:	Move production material to the site of the production processes as required
Trigger:	Delivery Schedule
Response:	All deliveries completed within 15 minutes of the scheduled time
Horizon:	15 minute, maximum
Outputs:	As2 - Production Material Ma2 - Production Material Ma3 - Production Material
Inputs:	MH3 - Stored Material
References:	MH4 - Delivery Schedule
Mechanisms:	None
Rules:	1. Emergency material moves from the receiving area take first priority 2. Emergency material moves from the warehouse take second priority 3. Damaged material is to be returned to the warehouse for corrective action before delivery

Figure 15–6. MH5—deliver material.

ANALYTICAL METHOD

The steps in the analytical method are shown in Figure 15–7.The first step is to discuss the output of subfunctions with the users, now that you know who they are. Ask them what they expect in terms of timeliness, quality, or usability, and cost if it is appropriate. If there is no conflict at this stage, note the rules by which you expect the

1. Verify timeliness of output
2. Verify usability of output
3. Question cost of output
4. Verify product responsibility
5. Document new resulting conflicts
6. Resolve all conflicts

Figure 15–7. Rules analytical method.

subfunction to accomplish its mission. Have you omitted anything that might be important or that might hamper performance if not considered?

If there is conflict at this stage, document it in the same manner as earlier conflict was documented. If there is disagreement among subfunctions as to who has the responsibility for the product, document that conflict. Now, to the extent the conflicts have not been readily resolved, it is time to "bite the bullet." There is no magic that the author can offer you at this stage. Look for common ground; look for possible compromise. The involved organizations should be the ones who try to resolve their mutual conflict. As they do so, there is the possibility that new conflict will be created with other units because of the resolution. Record this and solve it.

Sometimes an arbitrator will be needed if the parties lose perspective. Again, no magic—just hard work, teamwork for the good of the business.

COMMON ERRORS

A most serious error at this stage is the failure to identify the measures of the output from the user's point of view. This is like flying blind. The business that ignores its customers is a dying business.

A second error is the failure to seek concurrence from the user that the rules will ensure that the product will meet his or her needs.

If you are overlooking something, the user will most probably spot the omission. What do you have to gain by keeping your method of operation a secret?

CAUSES AND CURES

Our old enemy time can be a frequent cause of poor work at this stage. It takes time to talk to the user. It takes time to write down the rules. It takes more time to review them with the user. It takes a lot more time to work out compromises when there are conflicts. The team leader must enforce discipline to the extent he or she can. Up to now it has been good, clean fun. If you cannot resolve the conflicts, all the previous work has been wasted.

Another cause of error is a lack of willingness to involve other units in the resolution of conflict. This is often a reflection of a conflict that has been taken personally or that seems to point at past failure of a unit of the organization. Try to keep personalities out of the discussions, which is possible if the team leader is strong and the team members are intent on the task at hand.

Sloppy thinking is always a danger. The documentation of conflicts is a means of avoiding this. Each conflict is to be recorded in writing. Responsibility for solution is to be assigned and written down. The target date for resolution is to be recorded, and the solution is to be written down when it has been developed (see Figure 12–3). Keeping score helps motivate people to play the game properly. We do as we are measured. If it proves necessary, you might also write down the user's measures of performance for each output. Normally, this is not needed.

QUALITY CHECKLIST

A checklist to verify the quality of your work is shown in Figure 15–8. Are the rules practical and workable on a daily basis? If they are not, you must go back to the well. After all, any weakness in the work to this stage will certainly show up in daily operations.

Have both parties, supplier and user, agreed that the measures of output are fair and that the rules seem to ensure that the measures will be met? Are timeliness, quality, and cost per unit covered by the rules? This can be done in the rules themselves or in the trigger,

1. Workable?
2. Agreed to by users of output?
3. Cover timeliness, quality, cost?
4. Information flowchart updated?

Figure 15–8. Rules quality checklist.

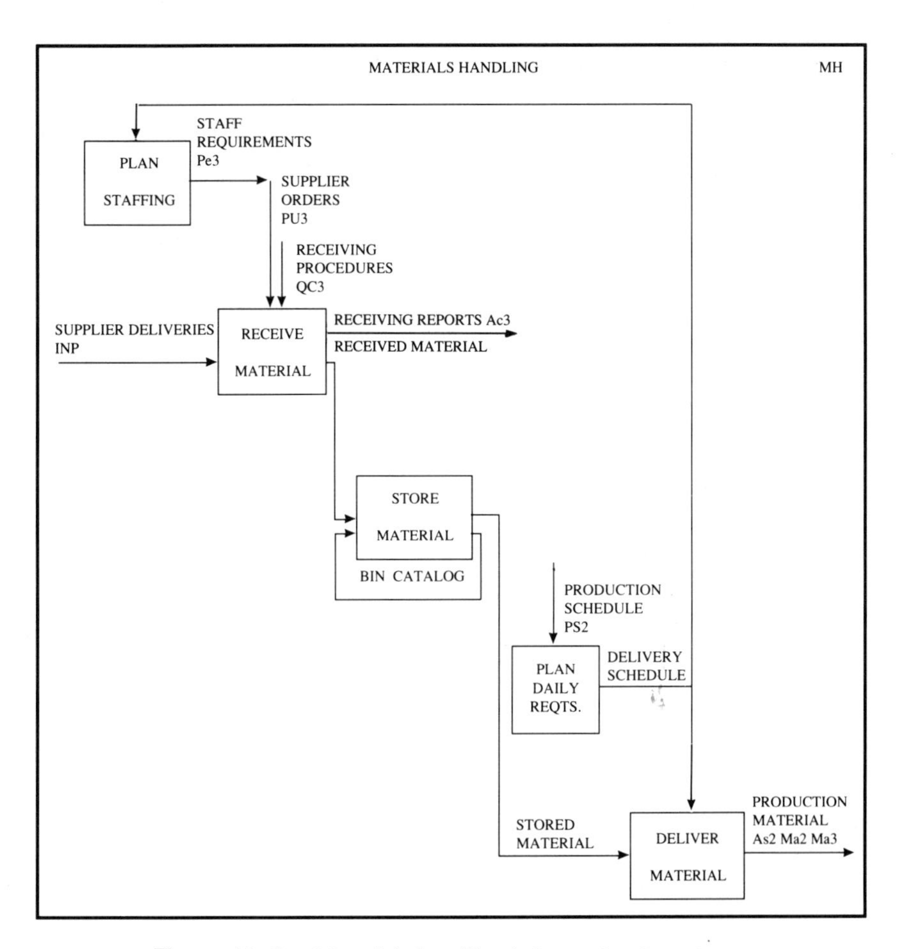

Figure 15–9. Materials handling information flowchart.

response, and horizon statements. Is cost reasonable, or will it later prove excessive, thus endangering the other measures of an output?

Have you updated the information flowchart with all changes in flow or subfunction? Note that in Figure 14–7 there was no known user for the receiving reports. In the resolution step, it was determined that Accounting needed these reports to authorize payment of suppliers. This resolution was recorded in Figure 15–3 and in the related flowchart as shown in Figure 15–9. A last check against flowcharts ensures that all areas of omission have been dealt with.

SUMMARY

This seems to be a simple topic, but it is the most difficult. The resolution of conflict by mutually agreed rules can be a complex and controversial task. The ability to perform well at this stage of FA is a function of the quality of the management and its willingness to deal with conflict. Keep in mind that conflict is a natural result of organization and the division of responsibility. It might help you through some hard times.

16

Computer Systems Requirements

INTRODUCTION

The time we live in is known as the age of the computer. The computer seems to be an intimate part of our life. It is pervasive. It is in the office, in the school, and even in the home. Our children are coming out of high school computer literate. Do we know what that means to us? Are we using the computer as much as we should? And are we using it as well and as wisely as we should?

The answer to the first question is no. The answer to the second is both yes and no. The answer to the third is usually no. Now why these answers?

We do not fully appreciate what the computer means to us. There are many reasons for this, but two should suffice here. There is so much "hype"about computers that it is hard to gain a clear perspective unless one works closely with them. And those who work with the computer often have such a love affair with it that they lose perspective on where the computer is best applied.

There is a tendency to automate because that is the thing to do. Most of the professional and trade journals are full of articles extolling the benefits of automation. Some types of automation have even been given special titles. When this happens, who is to speak against the trend? It is a brave manager who questions the ultimate wisdom of CIM, JIT, CAE, and so on. Are these modern applications of the computer wrong? No, they certainly are not. But they are not right for every business. Before you allow your actions to be forced by the current trends, it is prudent to do a little disciplined thinking.

The solution to most problems of information management seems all too obvious. Use a computer. Wrong! The use of functional analysis (FA) should reduce the volume and complexity of most of your information functions. Then you can ask which of the remaining functions really could benefit from the use of a computer. It is like any other tool: There are appropriate uses and inappropriate ones.

The exercise in this chapter will help you to reduce further the number of functions that appear to need automation. It will also be of some help in establishing priority among the functions competing for automation. The analysis technique shown in this chapter is not foolproof, but it can go a long way toward helping you to decide where to use a computer. Best of all, it does not require you to be computer literate in order to use it.

COMPUTER STRENGTHS

The computer has a few significant strong points that make it very useful. They are summarized in Figure 16–1. These specific characteristics of a computer are important to the management of information.

For most purposes, the computer processes data at a far greater speed than a person is capable of achieving. Consider an example that is out of this world, literally. If you are an astronaut and need to make a midcourse correction in a flight to the moon, you will benefit from use of a computer. In fact, you would most likely be past the moon before you could calculate the course correction manually. A computer, if it is accurately programmed and fed with accurate data, would be a valuable tool to an astronaut, since it could calculate the new course with great speed. This degree of speed is not often so vital in a business. But it is certainly needed for computer graphics and to manage the operation of complex machine tools.

A computer performs calculations with great accuracy. In over 30

1. Speed
2. Accuracy
3. Repeatability

Figure 16–1. Computer strengths.

years of professional experience, the author has seen only three times where a computer had an internal failure that caused an incorrect result. Note here that the author is not talking about how many faulty programs he has seen. If you were in a rocket, would you want the midcourse correction to be calculated with accuracy? Certainly you would. And you would want your paycheck to be calculated accurately. And you would want the material requirements of your business to be calculated accurately, as well as your bank balance.

Repeatability is a corollary of accuracv. An example is the ability to count votes multiple times and return the same result each time. Consistency is vital to many operations. It is certainly needed for the operation of an automated machine tool. Approximate accuracy is not acceptable in many computations. It is necessary that the calculating device always follow the same calculating "recipe" and always to calculate to the same degree of mathematical precision.

These three characteristics of a computer do not always make it the device of choice for all calculations or for all data processing tasks. It is of value where any other method is unacceptable in terms of performance and cost. It is of value where the product of the process needs to be of exceptionally high accuracy (quality), or where it is needed very quickly (timeliness). It is of value anywhere it can do the task at a lower cost per unit. So it is apparent that a computer performs a function just like any other business tool.

PERFORMANCE REQUIREMENTS

The above concepts deserve a bit more attention. You can evaluate the potential use of a computer for a given task by reviewing the performance requirements listed in Figure 16–2.

What is the response requirement of the task you are evaluating? How quickly is an answer (output, product) needed once the trigger event happens? The results of a calculation of the number of assemblers needed next month are usually not needed urgently. But you might be in a big hurry to find the location of an urgently needed batch of production material to feed a starved assembly line.

What is the accuracy requirement of the task you are evaluating? Can a degree of error be tolerated and yet not destroy the usefulness of the result? Is an approximate answer satisfactory? Calculating

1. Response Requirement
2. Accuracy Requirement
3. Complexity of Calculation
4. Volume of the Data Base
5. Volume Accessed
6. Frequency of Use
7. Record Update Frequency

Figure 16–2. Performance requirements.

the possible work load on the shop floor for the next five years requires only approximate accuracy. But you need a great deal of precision when you calculate the path of a cutter to make a part to precise tolerances.

What is the complexity of the calculation that is performed in the task you are evaluating? Is the calculation process of such complexity that manual calculation is likely to be error prone? The author once worked as an engineer and could never calculate the dimensions of a gear without making an error. The calculation was too tedious and required too many decimal places of accuracy. There were too many opportunities to make an error. Maybe this is why the Internal Revenue Service checks tax returns by computer.

What is the volume of the data base needed to perform the task? How many records are there in the data base? For the reader who is not computer literate, think of a data base as a file cabinet full of documents—records. Is your file a simple black book with a few dozen notations, or is it a massive reference file? The customer set of a manufacturer of military aircraft is smaller than that of an automobile manufacturer. For this reason, the evaluation of past sales is a simpler task for the aircraft manufacturer than it is for the automobile manufacturer.

What is the volume of data that must be accessed in order to perform the task? How many documents must be pulled from the file cabinet? How many records are accessed to perform a single calculation? Only one purchase order is needed to check the receipt of material from a supplier; but all bills of material are needed in order to calculate the material requirements for a factory.

What is the frequency with which the task is performed? How often is the action repeated? A materials storage system needs to look up the storage location of material many times in a day. But the assembly schedule of a manufacturer might not be calculated more often than once a month.

How often are the master records updated, if indeed there are master records associated with the task? That is, how often do you need to change or to add to the content of a typical document in your file cabinet? This is done every minute for the bin catalog of an automated warehouse, but it might be months between changes to a part drawing.

MEASURES OF DATA PROCESSING REQUIREMENTS

To apply the measures of performance requirements to your task, you need to put dimensions on them. Some example measures are shown in Figure 16–3. Each of these requirements can be roughly

	LOW	MEDIUM	HIGH	EXTREME
1. Response Requirement	>day	——	<min.	<5 sec.
2. Accuracy Requirement	10%	——	<1%	total
3. Calculation Complexity	arith	——	sort	iterate
4. Volume of the Data Base	<10	——	>1000	>100,000
5. Volume Accessed	<10	——	>100	>1000
6. Frequency of Use	>week	——	<hour	<min.
7. Record Update Frequency	>week	——	<hour	<min.

Figure 16–3. Measures of data processing requirements.

measured as to the degree to which it applies to a specific task. The measures apply to the output of the process, to the way the process is performed, and to the data that must be accessed while performing the task.

The broad measures are low, medium, high, and extreme. These very subjective words give a bit of a feel for the type of dimensions that might be required. Numbers are shown in Figure 16–3 for low, high, and extreme requirements. These numbers will apply to most tasks that you might consider. As you can see, the medium range is a very broad spectrum. It is left for you to apply dimensions to this measure as you wish.

To the extent a function has requirements for high or extreme performance in any of the seven parameters, then it is a likely candidate for some degree of computer support. If a function has high or extreme requirements in a number of the seven parameters, then it is a very likely candidate for application of a computer. The potential value from the use of a computer is amplified if the data are reused for other purposes.

The response requirement is low if the answer is not needed in less than a day, and it is extreme if it is needed in less than five seconds. The response requirement is high if it is needed in less than a minute. Of course, a person at a computer terminal is not likely to be willing to wait as long as a minute for the response to a transaction. Five seconds is usually a tolerable delay. A person working on a drawing at a graphics terminal is not likely to be satisfied unless most responses are less than a second. Graphics requirements will be dealt with in Figure 16–7.

The accuracy requirement is low if an error of 10 percent or more can be tolerated, and it is extreme if no inaccuracy can be tolerated. The accuracy requirement is high if less than 1 percent error can be tolerated. If you use a computer, any significant inaccuracy is more likely to be in the input data than in the computing process itself.

Calculation complexity is low if only simple addition or subtraction is involved, and it is high if it involves the sorting of large volumes of data. It is complex if iteration is involved. This is the case in expert systems and in the use of successive approximation to fit curves, as in regression analysis.

The volume of the data base is low if it has less than 10 records, and it is extreme if there are more than 100,000 records. The volume is high if there are at least 1000 records. Did you ever try to sort 1000 pieces of paper?

The volume of records needed to perform the task is low if less than 10 records are accessed in response to any single trigger. The volume is extreme if more than 1000 records are accessed. The volume is high if more than 100 records are accessed. Note that the explosion of a single bill of material involves from 5 to 50 records. The number of operations in a typical factory work order is from 5 to 30, which is why calculations based on the contents of bills of material and work orders are usually computerized.

The frequency of use of the function is low if it takes place less often than weekly, and it is extreme if it takes place at least every minute. The frequency is high if the task is done at least hourly.

The record update frequency is low if it happens less often than weekly, and it is extreme if it takes place at least every minute, usually with every response to a trigger. The frequency is high if it happens at least every hour. Consider an airline reservation desk where most often the performance of a task also updates the data base.

EXAMPLES

Two examples of how to evaluate a task are shown in Figures 16–4 and 16–5. The format is simple and leads to a reasonable picture of the possible need for application of a computer. Both examples are for Hydroclean, a company that has been used throughout this book.

An example of the bin catalog from Figure 15–4 is shown in Figure 16–4. There is no overwhelming need to automate the bin catalog, despite the extreme requirement for accuracy. The manual input of data to a computer could be as error prone as manual entry of bin locations on a master chart. But if the arriving material had a bar-coded bin tag for other purposes, the automation of the bin catalog could reduce the error potential from data input.

An example of the staff requirements calculation from Figure 15–2 is shown in Figure 16–5. There is obviously no need for mechanization of this process.

METHOD OF REPRESENTATION

The layout of the analysis as shown in Figures 16–4 and 16–5 is not rigid. Any organized method of recording the analysis should be satisfactory. You might want to consider some sort of weighting scheme such as 1 point for every medium measure, 2 for high, 3 for

BIN CATALOG

Subfunction: MH3 Store Material

Data Base: Bin Catalog

Response Requirement:	high	- <10 sec.
Accuracy Requirement:	extreme	- no error
Calculation Complexity:	low	- arithmetic
Volume of the Data Base:	medium	- 700 locations
Volume Accessed:	low	- 1 to 3
Frequency of Use:	medium	- <15 minutes
Record Update Frequency:	high	- <15 minutes
Other Data Base Uses:	none	

Figure 16–4. Bin catalog measures of performance.

extreme, and 1 for every reuse of the data base. But beware of becoming too automated in your analytical thinking. You could lead yourself into traps for lack of looking at the details and thinking about them.

ANALYTICAL METHOD

The steps in the analytical process are shown in Figure 16–6. The measurement system is merely an approximate scale, usable by the manager of the involved function. The first task is to estimate the requirements for each performance measure. Next you want to look for and count the number of high or extreme ratings for any of the seven requirements. If you find any extreme measures, they certainly suggest that you consider the use of automation. If you find three or more high measures, they tend to indicate that the task is a prime candidate for computer support.

STAFF REQUIREMENTS		
Subfunction: MH1 Plan Staffing		
Data Base: Delivery Schedule		
Response Requirement:	low	- 1 day
Accuracy Requirement:	medium	- within 5%
Calculation Complexity:	low	- arithmetic
Volume of the Data Base:	low	- 4 products
Volume Accessed:	low	- 4 products
Frequency of Use:	low	- monthly
Record Update Frequency:	—	- not applicable
Other Data Base Uses:	MH5 - Deliver Material	

Figure 16–5. Staff requirements measures of performance.

The notes as to reuse of the data base for other outputs can indicate possible candidates for automation. Each task might not merit automation by itself; but the full set of tasks that use the same data can possibly profit from use of a computer.

In all cases, it is well to make a rough estimate of the cost to automate and the potential cost avoidance or profit to be expected from automation. The final purpose of a business is to earn a profit. If you apply a computer where it yields cosmetic value but merely adds to the cost of operation, you have not acted prudently.

A data processor's participation is helpful, since he or she will be aware of the potential reuse of data and of the costs to automate. A first rough estimate of cost and value will tell you whether the subject deserves deeper study.

1. Estimate requirements
2. Count high and extreme ratings
3. Identify reuses of the data base
4. Evaluate costs and benefits

Figure 16–6. Computer systems requirements analytical method.

GRAPHICS APPLICATION

The requirements for graphics applications are much like those for data processing applications—but with some key differences due to the nature of graphics. These requirements are shown in the chart in Figure 16–7.

The preparation of a drawing, even with the help of a computer, usually takes more time than the processing of typical data transactions. A drawing frequently takes hours or even days to develop. For this reason, the response requirement scale is different from the scale for data processing. It does not relate so much to the time it takes to complete a drawing but to the time it takes to do a portion of a drawing. This could be placement of a line, copying of a portion of an image, or rotation of an image.

The amount of calculation needed to make a drawing is almost always high or extreme, so the measure of calculation complexity is not appropriate.

The volume of data in the data base refers to the number of drawings and not to conventional data records. It might refer to finished drawings that you will need to retrieve. It might also refer to standard images, such as bolts and bearings, that you want to copy from a library.

The two frequency measures are on a larger time scale for the same basic reason as the response requirement: They refer to the retrieval and use of a whole drawing and not to bits and pieces of a drawing.

	LOW	MEDIUM	HIGH	EXTREME
1. Response Requirement	>week	——	<day	<hour
2. Accuracy Requirement	10%	——	<1%	total
3. Volume of the Data Base	<10	——	>1000	>100,000
4. Frequency of Use	>year	——	<week	<day
5. Record Update Frequency	>year	——	<week	<day

Figure 16–7. Measures of graphics requirements.

TEXT PROCESSING APPLICATION

The requirements of a text or word processing task are similar to those for a graphics application. These are shown in Figure 16–8. The volume of the data base refers not to a number of records or sentences but to pages of completed text. The accuracy requirement really does not apply in this case, however, since there is no calculation.

	LOW	MEDIUM	HIGH	EXTREME
1. Response Requirement	>week	——	<day	<hour
2. Volume of the Data Base	<10	——	>1000	>100,000
3. Frequency of Use	>year	——	<week	<day
4. Record Update Frequency	>year	——	<week	<day

Figure 16–8. Measures of text requirements.

MACHINING SYSTEMS

The characteristics of a machining center and a manufacturing cell seem to be like those for conventional data processing. For this reason, the factors in Figure 16–3 apply quite well.

NEW TECHNOLOGY

If the task that you are evaluating seems to be a good candidate for automation but needs the use of a new technology, have a care. New technology brings risks you want to be sure you are prepared to take. New technology is by definition "new." It is not fully proven and can have bugs and unexpected side effects. Ask yourself what the alternatives are. Can you wait for the technology to mature? Will you risk a loss of competitive advantage? Are the costs well thought out?

And ask when you might need to standardize the new technology. The broad use of a new technology too early can lead to de facto standards you will have to live with for years, even after a better technology is made available. Of course, if you wait too long, the competition might jump ahead of you. It is a tough choice.

A good example of the problem with a new technology is Computer Aided Drawing (CAD). The engineering drawings made with an infant CAD package will have to be converted when you finally acquire a mature system. You must ask yourself if you will gain value from the use of the immature system. Can you reduce the cost of making drawings with the new technology? What might the costs of conversion be? Will the mature technology be in any way compatible with the infant technology? What is the value of gaining early experience? After all, you have to begin learning sometime.

SUMMARY

There is no magic formula for deciding what tasks to automate. This chapter contains one of many possible methods for evaluating the need and opportunities for use of a computer. It also suggests a way of prioritizing the opportunities. The concept applies to more than the processing of data. It applies just as well to evaluating the need for hard automation of physical processes, such as in a factory or in a print shop. There is a way of deciding when to use a text or graphics processor. There is a way of deciding when to use N/C (numerically controlled) machines and when to use robots. You can even use the technique to decide when to use automated manufacturing and assembly cells.

One important key to deciding when to automate is the degree of potential reuse of documents or records. The more frequent the reuse, the more likely that the application of a computer will be of value. Here is a critical thought to consider: If specific data must be available in mechanized form for one task, then the activity that originally produces the data is also a candidate for automation. Data are most accurate and timely if they are a natural by-product of the performance of a business function.

17

Computer Integrated Manufacturing Issues

INTRODUCTION

The introduction of Computer Integrated Manufacturing (CIM) is without real precedent in the field of computer aided processes. The first uses of numeric control and robots were as stand-alone technologies and were simple processes compared with the complexities of CIM. CIM brings with it the need to integrate the flow of material and information to an extent never before needed on the shop floor. And CIM in its broadest meaning also includes the automation of many office functions. It has an impact on the design of product and the design of production processes.

For these reasons, management has need for better tools with which to evaluate the potential use of the new technologies. The trade press seems to be telling management that CIM is the only way to go. The media say that CIM is the wave of the future. They imply that industry must automate or become obsolete.

Just to follow the all-encompassing recommendations of the trade press can lead to disaster if there is no businesslike way to measure the degree of benefit or of need. Functional analysis (FA) is a tool that can help to gain value from CIM while avoiding the risk of going too far in automating those things that should remain under manual control. FA was developed on the basis of experience in conventional data processing, but it appears to be just as applicable to evaluating the value of, and the need for, CIM.

CIM TOPICS

CIM is a label that has a very broad application. Many different disciplines seem to fall under this label. CIM includes CAD and CAM, the application of graphics technology to the design of products and production processes. CAD and CAM apply to the preparation of drawings of products and to the design of the tools and the processes by which products are made.

CIM includes use of JIT, or Just-in-Time, material procurement and production. JIT is based on the use of conventional data processing and might also make use of automated materials handling.

CIM includes the use of computer control for manufacturing systems and cells. This is called hard automation and involves the linked use of a number of modern technologies. These tools are things such as numeric control, robots, automated materials handling equipment, and coordinate measuring machines.

CIM includes the use of artificial intelligence in the form of decision-making algorithms in what is called an expert system. It also includes the ability to interpret images to control the movement of materials. It includes voice input and output for computer-driven processes.

FA is a valid technique for determining the degree to which each of the above technologies might be of value to your business. After all, each subfunction makes a product that can be tested by FA against the needs of the user. This test is as valid for materials and service as it is for information flow.

MATERIALS MANAGEMENT

How should you manage the delivery of production materials to the point of use? How are the materials planned for, ordered, received, moved? As was shown in Chapter 7, your choice of strategies can lead to simple practices or to an intimate dependency on the computer to drive the movement of materials. Simple practices are best, so long as they serve the needs of the business.

How will inventory record accuracy be maintained? How will the calculation of requirements be done? Are the past practices adequate? Do you need the output of these processes to be faster or more accurate? Are the triggers more frequent, as they would be from a manufacturing cell that is driven by JIT principles?

FA is a way to evaluate the demands each subfunction will place on other subfunctions.

PAYING THE BILLS

How do you account for the receipt of materials from a supplier? How do you handle erroneous deliveries or scrapped materials from a supplier? How do you pay the supplier's bill—based on each delivery or invoice or based on what the supplier should have delivered to meet the schedule?

With daily or even more frequent delivery of materials, the record keeping can become overwhelming unless you find a simple way to do it. JIT materials management makes it possible to place materials under close control. But JIT places more stringent demands on the information processes that it uses to plan, to order, and to track materials.

Again, FA is a way to evaluate the demands each subfunction will place on other subfunctions.

BILLS OF MATERIAL

How are the various bills of material managed and kept in line with one another? Is there a controlling bill of material or a master bill of material system?

Is there the need, or the ability, to tailor a bill to meet a specific department's need? If so, is there a way to keep it accurate when there are changes in the master product bill of material? Is there a way to keep the various bills in synchronism without adding or losing material?

How is product and process change identified and controlled? Have the demands on the bill of material system changed with the introduction of CIM? How are you to deal with the changes?

One more time, FA is a way to evaluate the demands each subfunction will place on other subfunctions.

PRODUCT CHANGE

Speaking of product change, how do you manage it in your business? Is there a reliable method for testing a change before it must be committed to? Is there a method for minimizing the need to scrap obsoleted materials?

How is product change reported and recorded? Do you have a reliable method of telling the customer the content of his or her specific unit of product? How does the customer know the components of the product when he or she has need to repair it?

SOMETHING ALWAYS GOES WRONG

A set of information systems and production systems can be a thing of beauty when all goes as planned. But what are the corrective and backup procedures when something goes wrong? The true test of the systems of a business is whether or not they provide an orderly means of dealing with the unexpected. And something always goes wrong. JIT materials management and production can become a disaster if there is frequent scrap or other breakdown of the production processes. Can the business still function with some degree of normality when the unexpected occurs? Or does everything become confused and fall out of order?

On the other hand, you want to be careful not to develop such successful backup procedures that people begin to accept failure as an expected norm. When you find the need to develop formal procedures for the rework of scrapped materials, you can be sure that your production process has a fatal flaw. There are many things that can go wrong:

- Customer orders can be canceled with no warning. And they will be if you take too long to fill them or have trouble meeting the customer's needs.
- A supplier can fail to deliver materials on time. This is the kiss of death to a JIT system, since it forces you to set aside an inventory of "just-in-case" materials.
- Or the supplier can fail to deliver usable materials for your production and assembly. Do you have a good way of filling in with other materials? Or, better yet, do you have a good way of ensuring that the supplier has tight control over his or her processes in the first place?
- You can be the victim of scrap from your own production processes. Do you have them under as tight a control as you expect of your suppliers?
- Lost material is always a possibility. This is most likely to happen in the delivery of materials from one location to another. Do you have a receiving system that tells you when

material has been delivered or tells you that the material is late and should be traced?

- The product can be flawed by any error in the assembly process. Do you have quality assurance steps in place for all processes? Is an error likely to be found and corrected before it is compounded in later assembly steps?
- If a design or manufacturing error is found in the field, when the product is in the customer's hands, it is much too late. You then have to undo many of the production and assembly steps in order to repair the damage. Worse yet, you have to ask how many other units of product you have produced with the same defect. And where are they? This is costly to correct, but the damage to your credibility in the eyes of the customer can be even more costly.
- In spite of the best-laid plans and systems, you are subject to the foibles of human nature. There most certainly will be staff shortages, strikes, and other impediments to production. How well are you protected from them?
- Your production processes can be shut down by equipment failures. This can happen to both computers and to production machines. Do you have a preventive maintenance plan that you follow to ensure equipment availability? Do you have backup plans or facilities, or can you tolerate the delay?

PRODUCTION MACHINERY

Do you have good systems to ensure that your production machinery stays in operating order? This question is worth asking a second time. The closer you come to JIT manufacture, the more dependent you are on equipment availability. The more you integrate the machines into manufacturing systems or cells, the more you are dependent on availability.

Do you need any surge tanks of parts to buffer for scrap or for equipment failure? If so, how do you check quality and make sure that the oldest material is used first? You certainly do not want to produce a large batch of material before you find out that it is defective. And you do not want material to degenerate over time for lack of use.

CELLULAR PRODUCTION

People often think of cellular production as the making of parts and assemblies one at a time, just as needed. This is rarely the case. Many parts need to be made in batches. This might be because they are used in batches. Or it might be because it is too expensive to move them to the point of use one piece at a time.

In your facility, what batch quantity does each cell make of each production part or subassembly? Is it a unit of one? That is not likely to be economical, since not all parts are used one at a time. An automobile uses four wheels at a time. It uses two of many other items, such as windshield wipers.

How are parts stored and moved between production cells? What is the economical lot size from a movement point of view? This is often called a transfer lot.

How does the mover know it is time to move specific material, how much to move, and where to deliver it? Is there any need to tell someone that it has been delivered? Whom do you tell, or what do you do, when you cannot deliver the material for some reason?

Do you have simple, self-governing systems for the management of production materials? Or do you have such a complex system that you need a computer to oversee it? Elegance comes from simplicity. The value of simplicity can easily override the hoped-for benefits of sophistication.

GRAPHICS

How are graphics images created and moved through the organization? Can everyone access a needed image with some degree of speed and ease? Are drawings developed from a library of common, simple images and indicators, or does every draftsman create his or her own images from scratch?

Is the design of a product done with conventions that make it easy to detail the parts? This is of service to Design Engineering itself. Is the part drawing done with conventions that make it easy to process the part? Is the drawing made in such a way that it helps the processor and tool designer to develop tooling and cutter paths? This is less likely, since it is a service to another department, Production Engineering.

How are changes in graphics images passed on to downstream activities? If some production processing is automated, how do you make the needed changes to the production process because of the change in the part that it produces? Is it done through incremental change to the production plan, or does it require total regeneration of the process?

Technology makes it possible for you to integrate the design and production planning work in your company. But common drawing and notation conventions are a key to success. They make it possible to reuse the work of "upstream" activities in order to reduce the effort in later activities.

Technology by itself is never a satisfactory answer. CAD is not magic. It must be accompanied by well-thought-out procedures to make use of the power of the system. These procedures must be the result of team effort of the various units of the business. Manufacturability is not the result of automation but the result of people who want to work as a team.

FA is a great aid to this process of working as a team. It is a way of developing systems that are integrated to smooth the flow of work.

DANGER OF AUTOMATION

There is a danger in the automation of production processes or the generation of product specifications by the use of a computer. To the extent that the processing rules have not been fully thought out, there is a degree of loss of control of the process. It is easy to crank out piles of garbage.

If the passage of time results in new requirements for an automated system, and it will, there is the possibility of finding out that you have developed very inflexible processes. Take the time to think out the implications of automation before taking action. Be careful to preserve in your people the skill and knowledge to perform each process manually. This will give you flexibility when you have the need to change. It will also ensure that your newly automated processes do not surprise you with bad results. Be careful also to determine the life expectancy and ROI (return on investment) of an investment in automation. Technology is evolving at an ever-increasing rate. You can find that you have need to change a process before you have earned back the money you invested in it. One can go broke chasing rainbows and looking for the pot of gold.

SUMMARY

This text does not propose to offer the answers to all the above questions relative to CIM. Each company must find its own answers. Each company is in some way unique from other companies.

What this text does hope to provide is a method, functional analysis. FA is a means for your company to find the answers to its questions about when and how to use the new computer-based technologies. It is hoped that through FA companies can identify the appropriate questions that they must ask of themselves. It is hoped that FA can help companies to find their own answers without a lot of pain and lost effort.

Do not mistake the intent of the author. He is not against automation and the use of the computer to improve quality and to reduce cost. Avoiding progress is a sure road to obsolescence and business failure. But blind acceptance of every new trend is an even faster way to business failure.

18
Summary

INTRODUCTION

This is a book about how to improve the procedures by which the day-to-day operation of a business is conducted. It is a book about information theory. Information is the coordinating medium that keeps the various components of the business operating in harmony.

This is not a book about computer theory, but it is vitally important to both computer people and to line managers. The reason for this is that it offers a method, functional analysis (FA), for a business to use to ensure that it has good systems.

PURPOSE OF THIS BOOK

The intent of this book is to encourage and to aid the practice of systems analysis for the business as a whole. How can we compete successfully with the Japanese, or with our domestic competitors, when we are competing within the business enterprise itself?

The book offers you a formal technique for systems analysis—Functional Analysis. It has proven to be very useful. It is hoped that you will find it to be of value to you.

FUNCTIONAL ANALYSIS

FA is a step-by-step approach to aid in the review and refinement, or restructuring, of business practices. By this means, it helps to integrate the resulting flow of information within the enterprise. It proceeds in a series of logical steps. These steps lead from corporate strategies to operating procedures. They lead to information systems and even, where appropriate, to the practical use of the computer. The technique urges the use of team discussion and problem solving among the various parts of the organization.

Chapters 3 to 16 of this book are each based on one of the steps of FA and are cookbooks on "how to" perform the analysis.

DO IT YOURSELF

The technique works for a subset of a business as well as for a business as a whole. It is hoped that you, the reader, can use it profitably in your own area of responsibility.

Ask yourself what your product is (or products are) and what your mission is. Ask yourself if your customer values your product and if you and he or she agree on the measures of performance of your mission, the measures of the product as needed by the user. Do you have any conflicts with your customers, with those who supply products to you, or even among the subdivisions within your own organization? How might you resolve these conflicts?

MISSION—PRODUCT—MEASURES—USER

The whole concept of FA is based on several simple premises. No unit of business exists except to perform a mission, which is to provide a product that is needed by the end customer or by another unit of the business. The performance of the mission is measured by three factors:

- The quality of the product—the degree to which the user finds that it suits his or her need
- The timeliness of delivery of the product to the user
- The cost per unit to provide the product.

The customer is the one who should determine the specific parameters by which the quality and timeliness of the product are to be measured. But often the supplier is unaware of the needs of the user—or simply ignores them. Sometimes the supplier is also a user and provides the product to serve his or her own needs first.

Because of need for cost control, and because of the division of functionality within the organization, there is a natural conflict that often occurs. If there is no transfer of cost, then the chain of command can place budgetary constraints on the supplier that degrade the product in the eyes of the user. If there is no transfer of cost from the user to the supplier, then the user can have unrealistic expectations of the product.

The purpose of FA is to provide an organized, constructive method for resolving these conflicts. The result should be local strategies and rules of behavior (procedures and systems) that minimize conflict and lead to the success of the whole business.

START SIMPLE

You, the reader, can use FA profitably in your own area of responsibility. Although FA is recommended for use by the business as a whole, why not start small and evaluate the area you know best, your own part of the company? In that way, you will also become more familiar with FA and will be able to apply it more broadly, with some assurance and credibility.

Solve your own problems before you worry too much about your neighbor's need to solve his or hers.

MANAGING THROUGH INFORMATION

People are often ignorant of their impact on their neighbors. People try to optimize their internal operation, often without thought about those whom they serve. It is often easier to operate that way, but it is much less effective.

A unit of business that operates with only its own self-interest in mind is apt to provide poor information for others. This leads to poor downstream operation and to the need to create redundant sources for information. Information is the lifeblood of a business. It flows through the units of the business, bringing them the means of knowing what they need to do, when it must be done, and where the output must be delivered. An integrated information system is a must—and FA is a way of obtaining it.

CAVEATS

The FA technique is sound and proven, but it is not easy to apply. It takes a lot of time, effort, and compromise. It does not yield instant results, but it does yield sound, lasting results.

FA requires open-minded, cooperative people on the analytical team.

There is the need for iteration, since something discovered in step 5 can cause you to redo some of the work from steps 2, 3, and 4.

Although FA is not easy to use, it is not confusing. It is quite logical as it progresses from step to step. It is not an abstract concept but a very practical one.

CIM APPLICATIONS

In its own way, Computer Integrated Manufacturing (CIM) is just another method of managing the information related to the production process. For this reason, it is subject to the disciplines of FA, just like conventional systems for information management.

OTHER USES

The results of FA form a compact and readable documentation of how a portion of a business operates. Once the analysis is complete, the set of documentation can serve as a basis for training new people in internal systems, the operating practices of the unit. The same documentation can also serve as a handy basis for reviewing practices when some important aspect of the business or environment has changed.

SUMMARY

The whole subject and purpose of this book is FA. It is based on the premise that information is a natural by-product of normal business activities. It is also based on a premise that conflict among business units is a natural by-product of organization. But conflict need not be destructive. It can be managed to improve the operation of the business. FA is thus a conflict management tool.

Best wishes for successful application of these concepts to the improved operation of *your business*.

Appendix

INTRODUCTION

There are 11 different types of documents that are developed during the course of a Functional Analysis (FA) study. Some of these documents are developed in stages during the various steps of FA. This appendix contains an example of each document with a bit of information about when it is developed

ORGANIZATION

The documents are gathered in four categories. They can be stored in a binder with four corresponding tabs for ready reference. The categories are:

- Business
- Department
- Conflict
- Computer

FUNCTIONAL ANALYSIS STEPS

The steps of FA are illustrated in Figure A–1 as a guide to when each type of document is developed. The documents are:

Business description — Figure A–2
- Developed in Chapter 3
- One page

Business mission—Figure A–3
- Developed in Chapter 4
- One page

Business goals—Figure A–4
- Developed in Chapters 5 and 6
- One page

Business strategies—Figure A–5
- Developed in Chapter 7
- One page

Business strategy rationale—Figure A–6
- Developed in Chapter 7
- One page per business strategy

Department description—Figure A–7
- Developed in Chapters 8 through 11
- One page per department

Department strategy rationale—Figure A–8
- Developed in Chapter 11
- One page per strategy per department

Department information flowchart—Figure A–9
- Developed in Chapters 13 through 14
- One page per department

Subfunction description—Figure A–10
- Developed in Chapters 14 and 15
- One page per subfunction per department

Conflict—Figure A–11
- Developed in Chapters 11 through 15 (described in Chapter 12)
- One page per conflict

Computer system requirements—Figure A–12
- Developed in Chapter 16
- One page per system evaluated

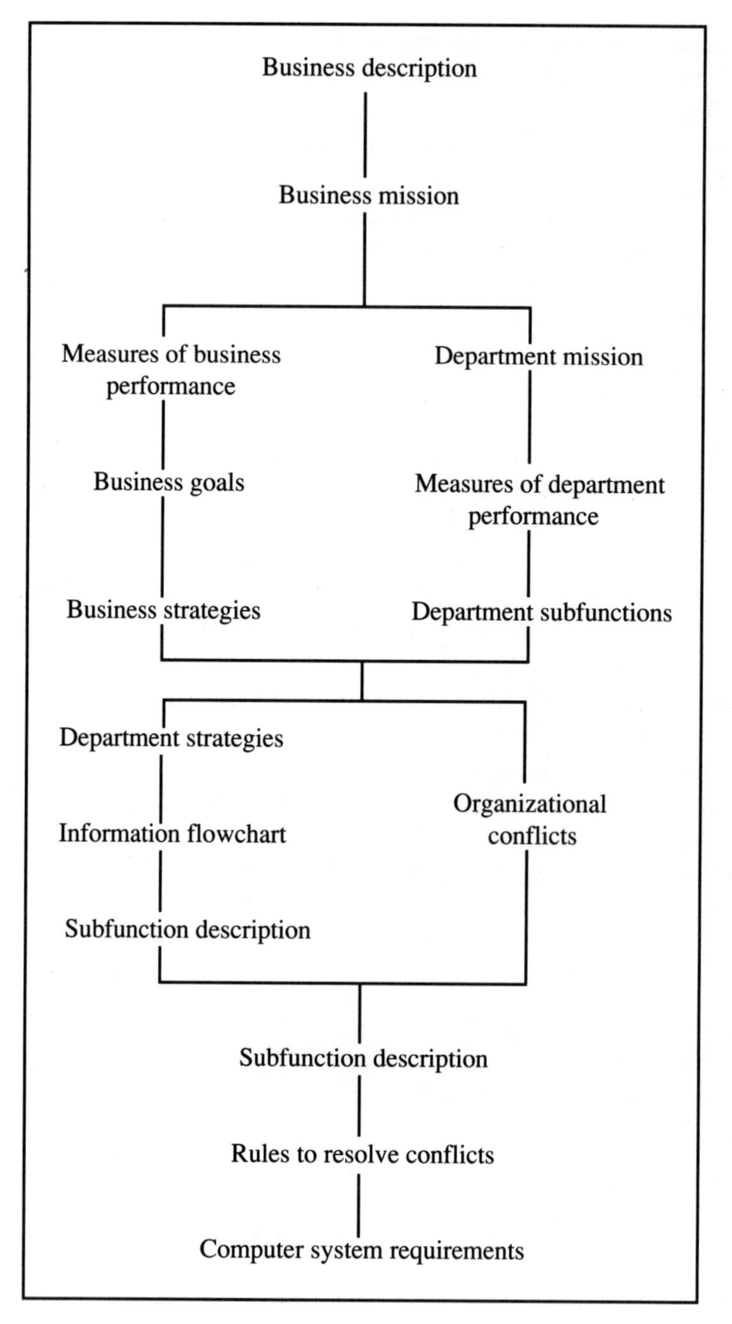

Figure A–1. Sequence of steps.

HYDROCLEAN

Description of company

* $200 million annual sales
* Manufacturer of washers and driers
* Located in the Midwest

Description of product

* Three models of each
* Four color schemes for each - baked on
* Retailer's label glued on when requested

Description of the marketplace

* Department store chains
* No direct retail sales

Physical facility

* Two assembly lines, each capable of handling both products
* Warehouse for storage of finished product
* Manufacturing area for frames, drums, and finish panels
* Subassembly area for control panels and heating units

Major operating practices

* 40-hour workweek
* Purchase of hardware, motors, and electronic controls
* Major change each model year
* Only safety or emergency change within a model year
* Part number change if a changed part is not interchangeable with the old part
* Product is built to stock and is sold from the warehouse

Figure A–2. Business description example.

HYDROCLEAN BUSINESS MISSION

Mission of the business

To market company-built washers and driers to retail chains

Business parameters

* The product is aimed at home use, not commercial use

* The product is sold under the label of the retail chain if requested

* The product is not sold in countries other than the United States

* Customer service is provided by the outlet, with training by Hydroclean

Figure A–3. Business mission example.

HYDROCLEAN BUSINESS GOALS

Owner viewpoint

A profit of 8 percent of sales

12 percent after tax return on equity

$300,000,000 annual sales in 1980 dollars

35 percent of total U.S. industry sales

Product viewpoint

Warranty less than 1 percent of annual sales

Less than 3 percent of products needing any sort of repair within one year of the sale date

95 percent of products shipped in time to meet customer-requested date

$100 enterprise cost per unit sold in 1980 dollars

$120 enterprise cost per unit sold in 1980 dollars - by 1995

Figure A–4. Business goals example.

HYDROCLEAN STRATEGIES

1. Prices will be within 5 percent of leading competitors.
2. Products will be aimed at home use.
3. The marketplace is the United States.
4. Distribution and product support will be through department store chains.
5. Product will be given the label of the department store chain when requested.
6. Frames, drums, panels, and wiring harnesses will be produced in-house.
7. Suppliers will be sole source for the items they supply.
8. Material procurement and production will be based on Just-in-Time principles.
9. Product change will be annual.
10. No new technology will be introduced until it has had two years of successful industrial use.
11. User instructions will be minimal and a part of the product itself.
12. Quality is targeted at ten years of life without repair from the time the unit is installed.

Figure A–5. Business strategies example.

HYDROCLEAN STRATEGY RATIONALE

1. Prices will be within 5 percent of leading competitors.

The intent of this strategy is to emphasize the strength of our company and the reputation of our product for quality, serviceability, and reliability. If the primary customer motivation for purchasing a Hydroclean washer or drier is price, we are put into a competition for cutting corners on those product characteristics that have differentiated us from the competition.

We do not intend, on the other hand, to pursue those customers who buy on the basis of high price or snob appeal. We will continue to emphasize the fact that our product can be installed and operated reliably for at least ten years from the time of purchase. Its continued reliable performance should be taken for granted by the customer, like a furnace or a telephone.

Because of the growth in apartment living and of utility rooms on the main floor of houses, the need for reliability and the avoidance of water overflow or leakage has become paramount.

By the same rationale, we must restrict our product outlets to chains with a high-quality image.

Figure A–6. Business strategy rationale example.

MANUFACTURING

Mission: To produce major components of Hydroclean products

Product: Major subassemblies

Customer: Assembly, Field Service

Measures: Less than 0.3 percent of production scrapped in production, at assembly, or in the first year of use

Less than 1 percent of assembly delay per week for lack of manufactured components

Cost per unit 10 percent less than delivered cost of purchased alternative

Subfunctions: Schedule Manufacture
Produce Components
Produce Assemblies

Strategies:
1. The manufacture of major subassemblies will be driven by the production schedule on a JIT basis.
2. Components of major subassemblies will be manufactured only if it can be done at 10 percent less cost than outside purchase.

Figure A–7. Department description example.

MANUFACTURING STRATEGY RATIONALE

1. The manufacture of major subassemblies will be driven by the production schedule on a JIT basis.

   Major subassemblies take up a large amount of space. The costs of storage and handling of subassemblies can be minimized if the production is tied directly to the schedule for assembly of final product, with material flowing directly from the subassembly process to the assembly line.

   With this direct flow, any quality defects in the subassemblies are more likely to be found and corrected.

   This JIT manufacture requires a material planning and procurement system that is also JIT. The possible disruption of the assembly process due to failure in the procurement or manufacturing processes is considered a necessary risk and a motivator toward prompt resolution of quality and delivery problems.

Figure A–8. Department strategy rationale example.

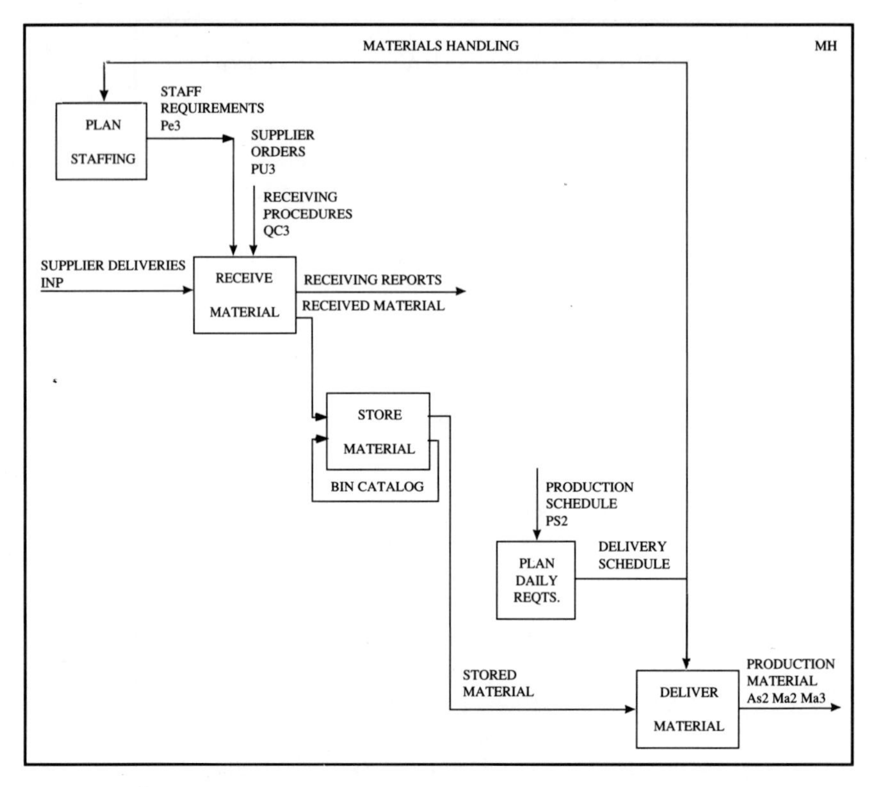

Figure A–9. Department information flowchart example.

PLAN DAILY REQUIREMENTS MH4

Function: Determine the schedule for regular delivery of production materials to meet the Production Schedule

Trigger: Receipt of Production Schedule

Response: Delivery Schedule calculated within one working day

Horizon: The Delivery Schedule is calculated for the horizon of the Production Schedule - the following calendar month

Outputs: MH5 - Delivery Schedule
*** - Delivery Schedule

Inputs: None

References: PS2 - Production Schedule

Mechanisms: None

Rules:

1. The delivery quantity is to be within the physical storage capacity of the using area

2. The Delivery Schedule is to smooth the delivery staff requirements as much as possible

3. 10 percent monthly staffing allowance is to be made for load consolidation and record verification

Figure A–10. Subfunction description example.

CONFLICT

Between: Manufacturing
Materials Handling

Subject: Delivery of production materials

Cause: Manufacturing JIT production strategy

Materials Handling strategy of overnight delivery of production materials

Solution: Storage at the manufacturing cell for a day's quantity of small materials

Hourly delivery of large materials

Figure A–11. Conflict example.

STAFF REQUIREMENTS

Subfunction: MH1 Plan Staffing

Data Base: Delivery Schedule

Response Requirement:	low	- 1 day
Accuracy Requirement:	medium	- within 5%
Calculation Complexity:	low	- arithmetic
Volume of the Data Base:	low	- 4 products
Volume Accessed:	low	- 4 products
Frequency of Use:	low	- monthly
Record Update Frequency:	—	- not applicable
Other Data Base Uses:	MH5 - Deliver Material	

Figure A–12. Computer system requirements example.

Index